Goethe's Art of Science

This book places Goethe's scientific method into context for 21st-century scientists and science. Moreover, *Goethe's Art of Science* is an excellent compendium for scientists interested in 18th-century and early 19th-century geology, colour theory and comparative biology.

Few scientists realise that Goethe contributed a significant body of work on geology, the theory of colour and comparative biology. Goethe's main concern was developing his scientific method called Morphology, which involved observing and describing form, whether it be minerals, rocks, the interaction of colour or organisms, the latter of which many taxonomists would be familiar with today. Goethe's scientific legacy was hampered by his Neptunist ideas in geology and his criticism of Newton on colour theory. Nevertheless, his view that our interaction as observers with natural phenomena was a fundamental principle that can be applied across many different scientific fields.

In order to raise awareness of Goethe's scientific method this book will:

- Provide a detailed description of his method, Morphology.
- Investigate his geological writings and how they were hampered by his Neptunist thinking.
- Demonstrate his Theory of Colours and how Anschauung and the Urphenomenon were crucial to his method.
- Follow the development of Goethe's plant and animal morphology.
- Explore what Goethe thought about his own scientific legacy and provide a list of his scientific achievements and honours.
- Provide new English translations of Goethe's writings.

Malte C. Ebach's research is primarily focused on Australasian biogeography, Gondwanan palaeobiogeography, natural classification (cladistics, biological taxonomy) and Goethe's scientific studies. He has written over 120 scientific articles, including 11 books, and a series of popular science articles and opinion pieces. In 2010, Malte and his co-author Lynne R. Parenti were recipients of the Smithsonian's Secretary Research Prize for the textbook *Comparative Biogeography*.

A volume in the Advances in Systematics Series

Series Editor: Manuel Lujan

Goethe's Art of Science

A Guide to Goethe's Scientific Method

Malte C. Ebach

CRC Press
Taylor & Francis Group
Boca Raton London New York

CRC Press is an imprint of the
Taylor & Francis Group, an **informa** business

Designed cover image: "Helix" by Julie Martin. Photo with permission Julie Martin.

First edition published 2027
by CRC Press
2385 NW Executive Center Drive, Suite 320, Boca Raton FL 33431

and by CRC Press
4 Park Square, Milton Park, Abingdon, Oxon, OX14 4RN

CRC Press is an imprint of Taylor & Francis Group, LLC

ISBN: 978-1-032-23038-2 (hbk)
ISBN: 978-1-032-23039-9 (pbk)
ISBN: 978-1-003-27534-3 (ebk)

DOI: 10.1201/9781003275343

Typeset in Times
by SPi Technologies India Pvt Ltd (Straive)

Dedication

Ἀφιέρωμα τοῖς ἀγεννήτοις καὶ τοῖς σχεδὸν ἀποθνήσκουσιν.

"You botanise and immerse yourself in optics? What are you doing?
Isn't it a better gain to stir a tender heart?
Oh tender hearts! A hack can move them;
Let this be my only joy, for you to be touched, by Nature!"

"Mit Botanik gibst du dich ab? mit Optik? Was tust du?
Ist es nicht schönrer Gewinn, rühren ein zärtliches Herz?
Ach, die zärtlichen Herzen! Ein Pfuscher vermag sie zu rühren;
Sei es mein einziges Glück, dich zu berühren, Natur!"

(WA I 1, p. 325)

Contents

Prologue

"Why Goethe?" is a question I'm often asked by colleagues, friends and family. "Wasn't he a poet?" Goethe was so much more than a poet. He was a polymath and a polyglot whose interests were unbounded, but one really consumed him like no other – Nature.

I too am obsessed with Nature and Goethe was a big part of my father's life. When I finished my doctorate and arrived in London as a postdoctoral fellow I rediscovered Goethe, but with a twist. He wasn't just a poet; he was a scientist. Ronald Brady introduced me to Goethe's ideas and to his large body of scientific writings. I was surprised by the overlap between what I studied and what Goethe wrote about Nature. I have a degree in geology, another in palaeontology and a third in biogeography, namely the study of the distribution of living (and fossil) organisms. I guess colour was the odd one out as I never really liked physics as a subject. Yet, Goethe's study of colour matched what my father did as a retoucher (what was done before computers and Photoshop): create, change and observe colour. I was stunned. How did this escape my attention? My father, a Goethe fanatic, never mentioned it and when I broached the topic with him, he seemed nonplussed. "He did geology?"

Goethe suddenly became an obsession, and I began reading his works. Not all his writings are translated into English, which posed no problem as German is my first language. Reading Goethe can be a struggle. The issue isn't the nuances of 18th- and early 19th-century German; it's just that Goethe's poetic tendencies could drag out a sentence over a paragraph, but in a superbly florid way. He doesn't get to the point quickly and, worst of all, he is quite repetitive. As Goethe scholar Rupprecht Matthaei tells it, "[he] has a tendency to tire the reader, if not scare him off completely". I stubbornly persisted until I published my first paper on Goethe's scientific work in 2005. What I noticed was that Goethe practised comparative science, which wasn't any different to what I or my colleagues did. That's why I have written this book – to show that Goethe was just practising comparative science (even in his work on colour). So why didn't anyone ever point this out? Probably because we don't publicise the fact. Taxonomic and comparative biology papers lack something that all scientific papers generally contain – a Methods section, which tells the reader what sort of instruments we used, be it a microscope or staining dye. What we don't tell you is what we are *doing*. We are observing, memorising organisms and their parts and recalling them. In effect, Goethe wrote the Methods section for taxonomy, comparative biology and colour science. Let me introduce you to these methods courtesy of this book.

This book is not a biography of Goethe's life, or how his life and writings have influenced others. There are biographies that deal with those subjects, such as the comprehensive biographies of Goethe by Nicholas Boyle, A. N. Wilson and Rüdiger Safranski. Rather, my book is a guide to Goethe's scientific method, and how he practised it and communicated his ideas in both his scientific and non-scientific writings. Similar books on Goethe's scientific writings were written by historians such as Karl J. Fink and George A. Wells, and scientists such as Agnes Arber, Wilhelm Troll and Dorothea Kuhn (botanists); Max Semper and Wolfgang von Engelhardt (geologists) and Rudolf Magnus and Rupprecht Matthaei (physiologists). Not surprisingly, it was practising scientists that thought Goethe's scientific method formed the basis for a comparative science, regardless of any trivial flaws it may have had. I aim to place Goethe's method back into the mainstream scientific canon where it belongs.

This book is aimed at scientists, so they can understand Goethe as a practitioner of science and how his science fits into practising comparative science today. I do not show Goethe's science as some forgotten or redundant attempt at Natural History, but rather as something that is still relevant. Goethe's science is still alive and well, and this book is dedicated to showing how it works, where it succeeds and where it falls apart.

I am grateful to David M. Williams for encouraging me to write this book. I thank Robin Bruce, Bernard Michaux, Evangelos Mantzios and Paul Gorjan for reading through earlier drafts. I also thank Julie Martin for her photography and editing, my wife Melinda Tursky for drawing the figures and Simone Meakin for editing the draft manuscript.

Malte C. Ebach
Wentworth Falls, Australia

A Note on Translation and Language

Every translation is an interpretation. Goethe's scientific writings are florid and repetitive, leading Rupprecht Matthaei to note that "Goethe's work has a tendency to tire the reader, if not scare him off completely". Also, Goethe used 18th-century German and terms that are no longer used in the same context. I have used *Grimms German Dictionary* [Deutsches Wörterbuch von Jacob Grimm und Wilhelm Grimm] to translate certain terms that have changed their use over time.

All translations are by me unless otherwise stated. I have included the original German for all translations using the abbreviations and attributions below. Translations from non-abbreviated sources will be given in full, and all original emphasis (e.g., italics, bold type), either in the original German or in the English translation, will be noted. Translations of texts in footnotes are not provided.

The original German sources used include:

WA = *Goethes Werke. Herausgegeben im Auftrag der Großherzogin Sophie von Sachsen.* Verlag Hermann Böhlaus Nachfolger, Weimar (1887–1919).

LA = *Die Schriften zur Naturwissenschaft, herausgegeben im Auftrage der Deutschen Akademie der Naturforscher (Leopoldina).* Kuhn D, Matthaei R, Troll W, Wolf L (Eds), Hermann Böhlaus Nachfolger, Weimar (1947–2012).

LA I 1, LA I 2 = On Geology und Mineralogy.
LA I 3, LA I 4, LA I 5, LA I 6, LA I 7 = On Colour: *Beiträge zur Optik* und *Farbenlehre*
LA I 8 = On Geology und Mineralogy (updated)
LA I 9 = *Zur Morphologie* (1817–1824)
LA I 10 = Essays, Fragments relating to studies in Morphology
LA I 11 = Essays, Fragments relating to studies in general
LA II = Supplements and Explanations to all volumes

Corpus = *Corpus der Goethezeichnungen* (7 volumes = I–III, IV_A, IV_B, V_A, V_B, VI_A, VI_B, VII). (Ed.) Gerhard Femmel. E. A. Seemann Buch- und Kunstverlag, Leipzig (1958–1973). I refer to volume, page and plate numbers (e.g., Corpus V_A, pl. V).

Faust I and *II* = I refer to Act and verse numbers as I have used various sources of *Faust I* and *II*.

Gespräche = Conversations with Eckermann and Soret are sourced from *Gespräche mit Goethe in den letzten Jahren seines Lebens 1823–1832.* Volume I (1836) and Volume II (1876) F.A. Brockhaus AG, Leipzig; and Volume III (1848) Heinrichshofen's Verlag Budhandlung, Magdeburg. I refer to names and dates only.

Deutsches Wörterbuch von Jacob Grimm und Wilhelm Grimm, digitalisierte Fassung im Wörterbuchnetz des Trier Center for Digital Humanities, https://www.woerterbuchnetz.de/DWB. I refer to webpages from this site.

I have used modern German spelling for any translations that have older 17th- and 18th-century terms that use the silent *h* in words such as *Thiere* [= Tiere] and *Charte* [= Karte], or the use of *y* in terms such as *seyn* [= sein], *bey* [= bei] or *drey* [= drei]. I have retained the use of the sharp-S (*Eszett*), namely *ß*, for both lowercase and capitals (i.e., "ss" and "SS"). Any technical or compound terms, specific to Goethe's own vocabulary, have been retained in the original German (i.e., *parallelpipedische*, *hindurchwirken*). Titles of texts are written in italics, whereas headings of texts are given in quotation marks. Translations or English titles after the original of names, texts or headings in texts are given in square brackets. Any German nouns used in the English text (e.g., Anschauung) are written with a capital as in the original German. Goethe will also be written without an umlaut (*contra* Göthe), even if written in the original text, and all original punctuation is retained. Also, original and translated titles of published and unpublished works are given in italics (e.g., *Metamorphosis of Plants*).

1 Goethe the Scientist

> What I have experienced, I know; and I consider experience to be the only true science.[1]

Johann Wolfgang von Goethe (1749–1832) is often portrayed by English speakers to be Germany's equivalent of Shakespeare, not unlike German speakers who will tell you that he was the greatest poet, dramatist and novelist in the German language. Few German speakers, however, would identify Goethe as a scientist. If they did, they would consider him a flawed naturalist who took on the mighty Isaac Newton (1643–1727) and thoroughly embarrassed himself.

Goethe's scientific writings are often misunderstood or misrepresented, if they are acknowledged at all. Regardless, Goethe was recognised by scientists[2] through the centuries as somebody who made a significant contribution to science.

> I am sure most of my readers honour Goethe only as a poet and a man; only a few have any conception of the high value of his scientific works, and of the gigantic stride with which he advanced before his own age – advanced so much that most naturalists of that time were unable to follow him.
>
> *(Haeckel 1876, p. 81)*

> Goethe's scientific work, which laid the foundation of comparative anatomy and plant and animal morphology, continues its fruitful influence to the present day.
>
> *(Magnus 1949, p. 124)*

> Among other things, [Goethe] was an accomplished botanist, he helped found the field of comparative anatomy, he coined the term *morphology*, and he anticipated the theory of evolution. If these achievements do not sound enough like "hard science," it is well to remember that in 1784 Goethe discovered the intermaxillary bone in the human jaw, thus supplying a link to primate anatomy that proved crucial to later evolutionary theories.
>
> *(Cantor 2004, p. 75)*

What did Goethe consider himself to be?

> On this moving globe the only joy and peace is found in true love, charity and the sciences[3]
>
> My mind drives me more than ever to natural science …[4]
>
> … strengthened through experience belief is again renewed, which is firmly based on the fact that nature has no secret, which it doesn't somewhere place naked before the eyes of the attentive observer.[5]

Goethe was a polymath: a poet, writer of screenplays and novels, an artist and a scientist, while at the same time working as an administrator dedicating his free time to all those pursuits. In modern parlance, Goethe was an amateur, living at a time when

DOI: 10.1201/9781003275343-1

science and art were becoming professionalised in institutions such as universities, art schools and museums. What we know of Goethe's scientific writings we have from published books, journal articles, letters and notes, which have been compiled, curated and published as *Die Schriften zur Naturwissenschaft, herausgegeben im Auftrage der Deutschen Akademie der Naturforscher* in 29 volumes over 70 years. Making sense of Goethe's writings is no mean feat and few have attempted to do so over the last 200 years or so (e.g., Magnus 1949; Wells 1978; Fink 1991). The problem is, and has been, how Goethe is interpreted. Many choose to interpret what Goethe says, his idle musings about natural processes or past events, which he neither returns to nor expands upon in any systematic way. Goethe did however do science in a systematic way and wrote about how one would practise it. He called this his Morphology, a universal method that is based on aesthetic appraisal of our observations and experiences we can recall from our memory (i.e., mind's eye). Goethe was not trained as a scientist; in fact, few people who practised anatomy or geology were educated in what we would now call the sciences. Yet, Goethe did understand historical scientific method (i.e., observe, record and classify), based on his work with Jena anatomist Justus Ferdinand Christian Loder (1753–1832). Goethe also understood the limitations of explanation, particularly within the realm of speculation.

> Imagination and wit [i.e., speculation], which, considered in isolation and applied to scattered objects, might be more dangerous than useful to a science, are nevertheless themselves the main tools with which genius reaches further than ordinary people are able to reach. If there are people who make very precise observations, others who organise and define what they have discovered [i.e., classify], and we must be very precise with the work of these people because they have taken on a very serious workload themselves, then we will take it all the easier with the third class, to which your friends confess for the time being, and they send you their warmest greetings.[6]
>
> *(LA I 11 pp. 31–32)*

Goethe had a good grasp of 18th- and early 19th-century science. Unfortunately, many compare Goethe's comparative science to experimental science (e.g., physics), leading many to consider the former to be "alternative" (Sepper 1988, p. 9), or something that "would not be recognisable to a contemporary scientist" (Hart 2024, p. 1). Both these assumptions are incorrect. Goethe uncovered what most scientists and artists do, namely aesthetic appraisal and classification, which is the most basic science we can do as humans, using our senses in the same way that an artist gauges the form of a human body, or a scientist describes a new species. Goethe was concerned with where the observer belongs in the scientific method. Clearly they are part of the scientific process, as it is the observer with their memory recall who compares and classifies the natural world. If so, why not mention our own experiences when we do our science? Why do we as scientists shy away from our own experiences when it is exactly these experiences, and our ability to recall them, that make comparative science possible? If we feel that our own observations and experiences are somehow flawed or subjective, then why do we even bother to go out into the world and find new species of organisms or minerals?

In Goethe's short poem *Parabasis* written around 1820, he declares "I am there to be astounded".[7] The polymath wants Nature to challenge him and reveal itself so he can observe and interact with it and be astounded. Trop (2013, p. 401, 404) summarised the poem as "empirical observation and the relation between the observing mind and the nature of that which is observed". The final line of the poem, Trop (2013) adds, highlights how "wonder connects and fuses the observer and the observed and makes them part of the same system". Goethe's focus on observation and the mind's eye is foundational to the scientific method that he had championed.

GOETHE'S SCIENTIFIC METHOD

Goethe's science of Morphology[8] is often, and correctly, identified as aesthetic in nature.

> Goethe's aesthetics was always so related to the empirical naturalism of his science ....
>
> *(Will 1956, p. 55)*

> It was his developing views on aesthetic appraisal that informed these epistemological precepts of his science.
>
> *(Steigerwald 2002, p. 303)*

What many of these positive critiques fail to mention is that aesthetic appraisal is also a *scientific* method, one that is practised in, for example, comparative biology, taxonomy and palaeontology. When we think of science, we think of repeatable experiments, statistics and analysis. The classic "hypothesis testing" enters our mind, where the scientist observes, questions, formulates a hypothesis and tests it again and again. The results are quantified and conclusions end up becoming scientific laws. While this method is common in physics, chemistry and medicine, it is virtually impossible to do in comparative biology.

If you were to discover a fossil, what questions could you ask other than "What is it?" In these cases you are simply left with your senses and reasoning. You describe the fossil, see if it resembles anything else you can observe. You compare it to other fossils. You compare it to living things. Once you have identified it as a plant, for example, you then make drawings or take photos. Along with your descriptions, you document it and compare it to other natural objects and then place it into a natural classification. You may have found a trilobite, a type of arthropod, which is a type of animal. Describing and classifying organisms is called taxonomy. In taxonomy there are no hypotheses, other than "is it a trilobite? If so, what sort of trilobite is it?", no repeatable experiments and no statistical inferences. Read a paper on taxonomy and you will discover something interesting. There is no Methods section.

Most scientific papers have four to five sections, including an Introduction, a Methods and Materials section, a Results and Discussion section, and a Conclusions section. No taxonomy paper has a true Methods section. That is to say, the author may mention what materials they used (e.g., a microscope, a pitfall trap, slides), but they never explain what they used to compare the organisms (e.g., observation and recall). An entomologist, for example, may compare hundreds, if not thousands, of specimens in a single study, but only produce a single image, be it a line drawing or photograph, of a single anatomical part or specimen. Where, then, is all that observational information of thousands of specimens kept? How is it accessed? No taxonomy paper will tell you. The reader is only left with printed images. If, for instance, an experimental scientist was to tell you that they looked at hundreds of specimens, they would be expected to show that data for each specimen examined, either in a table, or if larger, a database. Yet this does not exist in taxonomy because information is kept as memories that can be recalled at a later stage. So, then, what has this got to do with aesthetics? Go and watch an episode of *Antiques Roadshow*.

Most watch *Antiques Roadshow* for the vistas of British grand mansions or the head-spinning evaluations. Others watch it because of the experts. These evaluations are based solely around the ability of the expert, be it in glass, silverware or ceramics, to recall similar specimens in order to identify, date and value each teapot, carriage clock and oil painting. These evaluations are based on expertise not unlike that of a palaeontologist or botanist. The techniques employed to describe or revise a biological specimen are similar to those used to evaluate a piece of silverware or a watch. These techniques are aesthetic in nature and are inbuilt, that is, reside in our senses and memory. We all employ these techniques in our everyday lives. While many of us are not experts in entomology or Renaissance art, we still have enough experience of the world and the objects within it to make sense of it all. We would know the difference between say, salt and sugar, even though each is a white crystal. We can taste the difference between sweet and salty granules; we may even notice that sugar has larger grains than table salt. In fact, we are experts in our own spaces. We know where things are kept within our homes. We know what our friends and family look like and their mannerisms. We are aesthetes of our own personal worlds.

Taxonomists take at least four years to train on a single taxonomic group. To understand a group, you need to see as much variation and as many specimens as possible. Some taxonomists may spend years on a group simply because it has thousands of species. Mineralogists (mineral taxonomists) and petrologists (rock taxonomists) do the same. While there are slightly over 6100 mineral types[9] they are found in association with other minerals – think of two crystals growing into each other – meaning a mineralogist can not only identify and classify a mineral but can even tell you the type of geological setting in which it occurred. The same is true for a palaeontologist. An assemblage of different types of fossils will reveal its age, based on previous observations that have been recorded. Taxonomists, whether they are botanists, zoologists, mineralogists, petrologists or palaeontologists, all practise this aesthetic technique. Yet none ever mention it … until Goethe.

Goethe's scientific method is an aesthetic technique that taxonomists employ every day.[10] Goethe called it Morphology,[11] a term he first used in 1795. I will capitalise it throughout this book to distinguish a description of form from Goethe's methodological process. This book is called *Goethe's Art of Science* because Goethe understood that aesthetics is central to comparative science. Just as art experts, like those on *Antiques Roadshow* who use their aesthetic appreciation to identify an object's provenance, any comparison in science is based on individual and collective knowledge.

MORPHOLOGY: ANSCHAUUNG AND THE URPHENOMENON

> In Morphology we propose to establish a new science not because of its subject matter, which is already known, but because of its intention and method, which lends its principles their unique form and intention and method.[12]

Morphology was to be viewed "... as a theory [Lehre] in and of itself ...".[13] Goethe's usage of the term "theory" differs greatly from the modern definition, namely, a "hypothesis" or "supposition". Dennis Sepper argued that Goethe's theory is

> more akin to something suggested in the root meaning of the ancient Greek *theoria*[14] which was the activity of the spectator, a seeing and recognising, a sense also conveyed by the German Anschauung.
>
> *(Sepper 1988, p. 17)*

In the same passage Goethe emphasised that his Morphology intended to "portray rather than explain"[15] something that conflicts with his work on geology (see below). Morphology, initially intended to be a handmaiden to biology [Dienerin der Physiologie], came to encompass all sciences that allow observation and aesthetic appraisal, be it chemistry, the study of colour, geology or meteorology.

The new science of Morphology had two goals:

> partly to pursue the discovery of new subjects [Gegenstände], and partly to arrange these subjects more in conformity with nature and their own characteristics, eliminating all that is arbitrary insofar as possible.[16]

The "new subjects" Goethe refers to are new natural phenomena, which are classified by their own characteristics (i.e., their parts) rather than on some external criteria or their worth to society (i.e., industry). For example, the *Natural History* [*Histoire Naturelle, générale et particulière, avec la description du Cabinet du Roi*] of Georges Louis Leclerc, Comte de Buffon (1707–1788), had organised the mammals based on their use to human society[17] rather than by any innate characteristics.

Morphology rejects teleology, or the purposefulness of an organism or natural phenomenon. Goethe's *Morphology* not only recognises the role of perception in taxonomy and physiology, but also the human bias that pervades natural history, namely abstract ideas based on reasoning rather than observation. The science of Morphology sought to make observation a discipline, one in which natural phenomena and their parts, rather than our own views, determine our classification of Nature and its parts. If the natural phenomenon, such as an organism, is to take the centre stage than, say, a supposition or hypothesis, then the organism's *whole* structure becomes the object under study. Wholeness is where Goethe tackles observation as a two-part process, namely, our physical senses and the effects that they produce. These "fall into two parts which are not easily separated, i.e., into a physical part and a spiritual part [menschlichen Geist]". Goethe used the term "menschlichen Geist", which Sepper (1988, p. 59), in my view, mistranslated as "spiritual", when in fact "human mind" would be far more accurate.[18] A new translation would read "these fall into two parts which are not easily separated, namely a physical part and the human mind". Goethe saw that these were parts of the same whole. As physiologists, we place too much attention on the physical part, namely empirical observation, making this new science one-sided. Think of all those photographs taxonomists take to show all the different parts and their variations. What about the whole organism? To see the whole, we need the second part, the human mind, and only then will we be able to enhance our way of seeing [Vorstellungsart].

The adage "the whole is greater than the sum of its parts", which can be traced back to various authors, including Aristotle,[19] is often used to describe Goethe's idea of wholeness. Yet, Goethe's concept of wholeness is rather more internal (i.e., human consciousness) than external (i.e., empirical observation) as the next few thought exercises will show.

I own a cat. I have known my cat since he was a kitten. I have spent over 10 years with my cat and I have seen him grow and develop into a rather large cat. If I show you my cat, are you seeing the whole cat?

Let's say we have an unlimited supply of scientific equipment, anything from a hand lens to an MRI machine, to an electron microscope. We can take images of the cat's cells, neurons and internal digestive system. Effectively we can dissect the organism without hurting it. Can we then document the whole organism?

For the sake of the argument, let's say that we can clone the cat and dissect the organism at every point of its development. We can take an image of every cell, or every part of the organism as it grows, including every activity, be it sleeping, eating or playing. Have we documented the whole organism?

What we are doing is no different to the taxonomist. We are attempting to document the physiology of the organism. Unlike the taxonomist, we are looking at every aspect of the organism beyond what is needed for a formal diagnostic description. In the end we're left with a staggering amount of information; namely photos, images, graphs, micrographs and more data than one person can ever hope to understand. Have we captured the whole organism?

The answer to all these questions is simply "no". The organism is far too complex for any one person to understand. Moreover, we have only documented the parts we can image or measure. We have not captured the cat's dreams, or fears or joys. Yet, as the cat's owner, I have understood what the cat likes, and what causes it fear and joy. Yet, I am unable to document this satisfactorily in graphs and images. As a conscious being, I have a mind that remembers things, such as the cat's behaviour and what it used to look like, which I can recall at a moment's notice. All this lies inside me ready for recall. While I can imagine my cat doing things it never did, say, running through a puddle under the Eiffel Tower, I know that memory is fabricated. The same is true for a palaeontologist who could imagine the head of a bird on the body of a theropod. But our recall is based on our experiences, such as observation, hearing, touch, smell and so on. Even with all the images and scans and graphs of all my cat's parts, I still know it as a whole being because I have experienced it over its lifetime.

Wholeness, then, is not about how much information or data we can collect and quantify, rather it is about how much we experience a natural object. We may never fully understand the whole organism, but we are able to work towards it as a goal. We seek to portray rather than to explain. Wholeness is the hypothesis that we wish to test. By observing and recalling we understand the object more than if we were to propose a hypothesis and then seek to demonstrate whether it is true or not. Understanding wholeness is the most fundamental aspect of science. To ask "what is that?" is the fundamental question. That is Goethe's Morphology, the aim to understand wholeness through using our senses and our mind, an activity that Goethe called Anschauung and the Urphenomenon.

Anschauung is that most basic skill, not "to see" or "look", but to *observe*. Together with our Mind's eye[20] the Urphenomenon is a result of an activity. In fact, Goethe refers to Anschauung as though it is a verb,

> When I finally rest from [the] Urphenomen[on]izing ....[21]

Any of our senses is Anschauung, be it sight, hearing, smell, touch or taste. We do not make much of these other senses in science, but these are important in, say, the arts, cooking and manufacturing, particularly when we place great value on them for their accuracy. A disciplined sommelier is *trained* and can empirically identify different types of wine based on taste, smell and experience. Notice my emphasis on trained. Observation, as with all senses, can be trained through practise and mentoring. Taxonomists need to be trained by other taxonomists, to get "their eye in" on certain parts. The same is true with any other profession that uses Anschauung. Simply "to look" or "to see" is not the same as trained observation, a skill long in the learning.

Ronald Brady (1937–2003) had this anecdote that clearly demonstrates poor observation.

> I once experienced my own lack of focus in this department in a particularly disturbing way. On long drives, dinner in less than satisfactory places becomes inevitable. After such a meal I made for the door with hopes of finally escaping the plastic plants that stood throughout the restaurant. As I arrived at the door and opened it on the parking lot, I realised that the large plant near the door, which I had seen as artificial, was real. I watched it "come alive" as my recognition set in. I felt accused of misprision by the plant.
>
> *(Brady 2001, p. 62)*

Noticing a plant, perhaps lurking in a dark corner, is not giving it your full attention. Suddenly, by giving it more than just a cursory glance, the plant, immediately recalled from an earlier passing, is transformed. A correction makes the plant appear in a whole form. The aim of attaining greater awareness of the whole plant has been achieved. What, then, is the whole plant? Here Goethe's thinking becomes quite disarticulated, as we shall see later, but through no fault of his own.

Goethe explained his view of wholeness as follows:

> The members of every creature are formed so that it may enjoy its existence, and maintain and propagate itself; in this sense everything is alive deserves to be called [vollkommen].[22]

Vollkommen is a fascinating term as it has various definitions, namely "perfect", "ideal" or "fully" as in the "perfect score", an "ideal life" or "fully developed". Yet

Goethe uses the term in the context of "complete", as in the "complete organism". Goethe continues,

> I would therefore like to say: we call a completely [vollkommen] organised being beautiful if, when we look at it, *we can imagine that it is capable of making a manifold [mannigfaltiger] and free use of all its limbs whenever it wishes*; the highest feeling of beauty is therefore linked to the feeling of trust and hope.[23]

Note the terms "manifold" and "free use" in relationship to "completely". If, for example, we use the term "perfectly", we imply imperfection, a purely subjective term. Yet "completeness" is objective, as we will know if an organism is missing a limb. A three-legged dog, however, is an incomplete organism, but to its adoring owner it is perfect. Can perfection have manifolds? I do not think it can, as the term only focuses on what is perfect and what is not, yet we can have varying degrees of completeness. But Goethe did not see completeness varying in degrees either. Perhaps perfection is too subjective, focusing on likes and dislikes, and completeness too mechanical, focusing on degrees or percentages. Goethe saw the organism as perfect/complete regardless of how many limbs were missing or by the degree of development (e.g., embryo is an incomplete adult).

The early 20th-century English botanist Agnes Arber (1879–1960) also argued this point:

> The quality, for example, of any animal or plant may approximate to perfection *in its own kind*; but these perfections do not lead on to one another, and they cannot be ranged along a scale. Each is ultimate in the sense of being an expression of the Absolute conditioned by the characteristics and limitations of a finite individuality.
>
> *(Arber 1957, p. 7)*

Arber wrote the above in her book *The Manifold and the One* in which she concludes that "… the One is the Manifold, and the Manifold is the One" (Arber 1957, p. 118). Her conclusion fits with Goethe's idea of a "complete" organism "which is capable of making a manifold". Perhaps the term "wholeness" or "die Ganzheit" would be a better compromise than "complete". If we use the term "wholeness" then "manifold" would make sense as it refers to form when used in relation to organisms. In this sense it makes Goethe easier to understand when we translate "vollkommen" as "wholeness",[24] "mannigflatiger" as "manifold" and "call a wholly organised being beautiful if, when we look at it, *we can imagine that it is capable of making a manifold and free use of all its limbs whenever it wishes*".

The Anschauung of an organism and its parts means we are able to see it as a whole over time: from when it is a seed, to it sprouting, forming a stem and leaves, flowering and then bearing fruit. Given it is impossible to see that at one go, say via a photograph or montage, it is possible to replay those experiences in your mind. We can take these experiences and play them in our minds, as though you are winding them forward or backward. The sceptic may say that these are imagined and not of any scientific value. Yet, this is exactly what a taxonomist does when comparing their experiences of organisms. A disciplined taxonomist is trained to observe and compare, meaning that their minds recall to the best of their ability accurate experiences of organisms and their parts. Of course, mistakes are made and anomalies may occur. But a trained taxonomist is unlikely to recall an anatomically correct leprechaun based on an actual experience of a living example.

The ability to recall and see our experiences of form based on Anschauung can be recreated in our mind's eye; this is what Goethe called the Urphenomenon.[25] Along with Anschauung, it is a basic tool of Goethe's Morphology. Goethe didn't discover Anschauung and the Urphenomenon; he just decided to make people aware of what comparative scientists and artists were doing as an active sensory and mental activity.

Morphology is a scientific method based on phenomena, be it an organism, light, colour, rocks, minerals or clouds. Each natural phenomenon can be experienced through Anschauung. Those experiences are recalled in our mind's eye as the Urphenomenon. These natural objects themselves are whole through our observations, and wholeness (an abstract idea) comes about when we perceive them as Urphenomena. In this sense the logical and historical sequence to Goethe's Morphology is:

$$\text{Morphology} = \text{Wholeness}\left(\text{Hypothesis}\right) + \text{Anschauung} + \text{Urphenomena}\left(\text{Practise}\right).$$

Goethe's Morphology developed between 1794 and 1799, well before the publication of his *Farbenlehre* [*Theory of Colours*] in 1810 and his *Zur Morphologie* [*On Morphology*] in 1817. In fact, much of his Morphology was written and developed well before it was published. For example, the terms Anschauung and Urphenomenon first appear in *Farbenlehre* (LA I 4, pp. 70–71), yet they are first described in 1795 and 1798, respectively. Anschauung first appeared in *Outline for a general Introduction to Comparative Anatomy, Commencing with Osteology*[26] in 1795 and was published in 1820, as well as in *Studies for a Physiology of Plants*[27] and *Observation on Morphology in General*[28] in 1795, both of which were published in 1891. The Urphenomenon had a similar delay. The Urphenomenon (termed "pure phenomenon") was defined and developed as a concept in his essay *Empirical Observation and Science*, written in 1798 but published posthumously in 1893.

The history of Goethe's methods of Anschauung and the Urphenomenon is interesting because much of Anschauung was developed from plant and animal studies, whereas the Urphenomenon was developed as a more general concept within optics (and referred to in his *Zur Morphologie* as the Typus (see Ebach 2005). Along with Goethe's work on geology, in which he was concerned that his Morphology was flailing, we can be assured that Morphology, with Anschauung and Urphenomenon as its underlying activities, developed primarily out of Goethe's work on colour, botany and zoology, showing that Morphology, as a method, is not confined to biology.[29]

Goethe's most notable treatise, *Beiträge zur Optik* [*Contributions to Optics*], was a primer towards his Morphological method, "[i]t has been my desire to apply the same expertise and conscientious effort here [*Beiträge*] as I have applied to botany [*Metamorphosis of Plants*]".[30] These works detailed what the observer was experiencing, thereby later allowing Goethe to describe the activity of the observer.

After all, Goethe's Morphology is the study of how we observe, measure and recall our experiences. Goethe clearly described these activities after his work on *Versuch die Metamorphose* der *Pflanzen zur erklären* [*Metamorphosis of Plants*] and *Beiträge zur Beiträge*, without which he would not have had the opportunity to experience firsthand what Anschauung and the Urphenomenon were.

GOETHE ON SCIENCE

This book investigates the role of Goethe's Morphology in his scientific work, where Morphology was successful and where Goethe failed to engage in Anschauung and Urphenomenon. Morphology was to be, for Goethe at least, a unifying method, one that encompassed art as well as science. Yet, when theories (i.e., opinions and models) get in the way, our ability to observe and experience phenomena directly is compromised. For example, Goethe was a victim of his own ideology when he based his ideas about the formation of granite and basalt on Neptunistic theory. His application of Morphology to the study of colour and plant and animal anatomy is much more consistent. This book is divided into seven chapters, not all of equal length. Goethe spent most of his time in Weimar working in the court of the Grand Duke Karl August, Duke of Saxe-Weimar-Eisenach (1757–1828) in various governmental roles. His role as inspector of mines meant he spent much of his time engaged in looking at rocks and dealing with mines and their construction, therefore many of his writings, mostly unpublished, are on geology. Goethe also spent a large part of his scientific studies on the theory of colour, which makes up almost half of his scientific writing. His work on biology and geography is far less comprehensive, even though these two areas contained a lot of his ideas, such as metamorphosis and the Typus. The size of each chapter reflects the work done in these areas. Geology and colour are the largest, while botany and zoology are the smallest.

Chapter 1, "Goethe the Scientist", is an introductory chapter to *Goethe's Art of Science*. It explains the aims of this book, namely to investigate Goethe's method, Morphology and the various techniques employed, such as the idea of wholeness, Anschauung as a form of observation and the Urphenomenon as a form of memory recall. I will also introduce the various fields of science where Goethe had sought to employ his Morphology, namely geology, colour theory, botany and zoology.

Chapter 2, "Goethe on Geology", the largest chapter, is an exhaustive account of Goethe's attempts to employ his scientific method in the Earth sciences. The start of Goethe's journey into the Earth sciences and mining started quite early during his time at the University of Strasbourg. Coincidentally his first job was within the ministry of mining in Weimar. Very early on Goethe adopted Werner's Neptunism, an idea that he stubbornly stuck with for the remainder of his life. Goethe's interpretation of Neptunism is explored in detail, together with the concept of the Urgestein and the formation of granite and the weathering of granite tors. The formation of basalt is also explored, as are Goethe's ideas on the formation of volcanoes, particularly the Kammerberg and his experiments on melting rocks and minerals. I also explore Goethe's views on the Carlsbad twinning in K-feldspars, on the naming of Goethite and granulite, on Goethe's ideas of glaciation and the Ice Age, his interpretation of meteorites and his work on the 1821 geological map of Germany.

Chapter 3, "Goethe on Colour", focuses on *Beiträge zur Optik* and *Farbenlehre* and other supplementary works. Goethe's concept of theory is established as being an activity rather than constructing passive arguments. The activities in the *Beiträge* are explained as a form of learning-by-doing. The development of his method between the *Beiträge* and *Farbenlehre* is explored, as is the development of the pure phenomenon, which he later called the Urphenomenon. The classification of colour is discussed as being the main thrust of Goethe's method in the *Farbenlehre*. The issue Goethe had with Newton's method (the polemic part of his *Farbenlehre*) and the historical section are briefly touched upon.

Chapter 4, "Goethe on Botany", looks at the development of the Urpflanze as being the hypothesis of a whole plant, an idea that via Anschauung leads to the development of the Urphenomenon. I also look at the artistic depictions of the Urpflanze that have captured the imagination of many artists and scientists since Goethe's day. The *Metamorphosis of Plants* is compared to the *Beiträge* as a learning-by-doing activity that involves Anschauung.

Chapter 5, "Goethe on Zoology", looks at the works published in *Zur Morphologie*, in particular the Typus and its role in descriptive taxonomy and comparative anatomy. The Typus isn't, as commonly believed, an archetype, but rather an outline of anatomical parts from which different organisms, such as mammals, can be systematically compared. I show that the Typus is still used in 21st-century taxonomy. The Urtier is compared to the Urpflanze as a hypothesis of the whole animal. Goethe's brief attempt at a Metamorphosis of Animals is also discussed.

Chapter 6, "Reimagining Goethe's Art of Science", attempts to reconcile all the concepts within Goethe's Morphology based on what he practised, rather than what he thought. In this chapter I propose to bring Goethe back as a practitioner in mainstream science, rather than someone who created his own alternative science that died with him. I reinterpret Goethe's Morphology from a modern scientific perspective. I also look at the terms used to ridicule Goethe's work and how these terms don't reflect what he did and discuss why they are incorrect. Goethe's works and method are also subjected to an analysis that shows that Goethe's writings only make sense when read in order of composition rather than publication. I look at Goethe's failures and his enduring recognition during his lifetime and how they led to his legacy.

Chapter 7, "Goethean Encyclopaedia", is all about Goethe's legacy, the minerals, organisms, mountains and other natural formations that are named after him. I look at natural curiosities that have been named after Goethe, including the Goethe Elephant and the Goethe Palm. Goethe's own view of his legacy is discussed.

NOTES

1 "Was ich erfahren habe, das weiß ich; und halte die Erfahrung für die einzige echte Wissenschaft" (Goethe to Friederike Oeser, 8 April 1769; WA IV 1, p. 206).

2 In his article *What does comparative anatomy owe to Goethe's scientific work?* [*Was verdankt die vergleichend-anatomische Wissenschaft der Arbeiten Goethes?*] Wilhelm Lubosch (1875–1938) notes "...what comparative anatomy owes to Goethe would be learned from studying these forty authors" ["... was die vergleichende Anatomie Goethe zu verdanken habe, der würde nach dem Studium dieser vierzig Autoren eines anderen belehrt sein"] (Lubosch 1919, p. 157).

3 "Auf diesem beweglichen Erdball ist doch nur in der wahren Liebe, der Wohltätigkeit und den Wissenschaften die einzige Freude un Ruhe" (Goethe to Charlotte von Stein, 1775–1776; WA IV 7, p. 269).

4 "Mein Gemüt treibt mich mehr als jemals zur Naturwissenschaft ..." (Goethe to Knebel 9 July 1790; WA IV 9, p. 213).

5 "... durch Erfahrung bestärkten Glauben wieder auffrischte, welcher sich fest darauf begründet, daß die Natur kein Geheimnis habe, was sie nicht irgendwo dem aufmerksamen Beobachter nackt vor die Augen stellt" (LA II 1A, p. 439).

6 Einbildungskraft und Witz, welche, abgesondert betrachtet und auf zerstreute Gegenstände angewandt, einer Wissenschaft mehr gefährlich als nützlich sein möchten, sind doch selbst die Hauptwerkzeuge, womit das Genie weiter reicht als gewöhnlich die Menschen zu reichen vermögen. Wenn es also Männer gibt die recht genaue Beobachtungen machen, andere, welche das Erkannte ordnen und bestimmen, und wir es mit den Arbeiten dieser Männer sehr genau nehmen müssen, weil sie selbst ein sehr ernsthaftes Pensum übernommen haben: so wollen wir es mit der dritten Klasse desto leichter nehmen, zu welcher sich vorerst Ihre Freunde bekennen, die Ihnen zusammen einen herzlichen Gruß entbieten. (LA I 11 pp. 31–32; Also summaries in Sepper 2009, p. 264)

7 Parabase

Freudig war, vor vielen Jahren,
Eifrig so der Geist bestrebt,
Zu erforschen, zu erfahren,
Wie Natur im Schaffen lebt.
Und es ist das ewig Eine,
Das sich vielfach offenbart:
Klein das Große, groß das Kleine,
Alles nach der eignen Art;
Immer wechselnd, fest sich haltend,
Nah und fern und fern und nah,
So gestaltend, umgestaltend –
Zum Erstaunen bin ich da.

8 Goethe's Morphology is distinguished from his 1817–1824 work *Zur Morphologie.*

9 The International Mineralogical Association (IMA) Commission on New Minerals, Nomenclature and Classification recognises 6213 valid species of minerals as of March 2026. https://cnmnc.units.it.

10 I will not make a big deal of it, but Goethe's method as employed by Goethe was not and is not an "alternative science". *Morphology* is practised by anyone looking at form. I wish to distinguish Goethe's *Morphology* from Goethean science, or *Goetheanistische Naturwissenschaft*, as championed by Wolfgang Schad, Jochen Bockemühl and Andreas Suchantke, has its origins in the work of Rudolf Steiner and his interpretation of scientific practise.

11 Steigerwald (2002, p. 318) correctly states that German physiologist Karl Friedrich Burdach (1776–1847) was the first to use the term "morphology" in print (as a footnote) in his *Propädeutik zum Studium der gesamten Heilkunst* [*Introduction to the Study of Medicine*] (Burdach 1800, p. 62). In the same volume Burdach also coins the term "biology" in a footnote (Burdach 1800, p. 62).

12 Indem wir in der Morphologie eine neue Wissenschaft aufzustellen gedenken, zwar nicht dem Gegenstande nach, denn derselbe ist bekannt, sondern der Ansicht und der Methode nach welche sowohl der Lehre selbst eine eigne Gestalt geben muß als ihr auch gegen andre Wissenschaften ihren Platz anzuweisen hat .... (LA I 10, p. 140; see Miller in Goethe 1995, p. 57)

13 "Die Morphologie kann als eine Lehre für sich ..." from *Observation on Morphology in General* [*Betrachtung über Morphologie überhaupt*] (LA I 10, p. 140; translation by Douglas Miller in Goethe 1995, p. 57).

14 *Theoria* or *θεωρία* may also be translated as "action of viewing, contemplation" OED (2025) https://doi.org/10.1093/OED/5775040472.

15 "Dar sie nur darstellen und nicht erklären will ..." (LA I 10, p. 140; translation by Douglas Miller in Goethe 1995, p. 57).

16 "... ihre Arbeit ist doppelt teils immer neue Gegenstände aufzufinden, teils die Gegenstände immer mehr der Natur und den Eigenschaften gemäß zu ordnen und alle Willkür, in so fern es möglich wäre, zu verbannen" (LA I 10, p. 141; translation by Douglas Miller in Goethe 1995, p. 57).

17 Buffon organised domestic mammals as so: the horse, the ass (donkey), the ox, the sheep, the goat, the swine and the dog.

18 A similar term "Geistesstimmung" or "state of mind" (LA I 11, p. 39) was in my view mistranslated by Herb Aach (in Matthaei in Goethe 1971, p. 65) as "soul" [in German "Seele"]. Douglas Miller (in Goethe 1995, p. 24) translated "Geistesstimmung" as "state of his senses", which is more in line with Goethe's intended meaning.

19 "But as for the question of the definition and the number, what is the reason for the one? For everything that has many parts is not a single whole, but rather is the whole something besides the parts" [περὶ δὲ τῆς ἀπορίας τῆς εἰρημένης περί τε τοὺς ὁρισμοὺς καὶ περὶ τοὺς ἀριθμούς, τί αἴτιον τοῦ ἓν εἶναι; πάντων γὰρ ὅσα πλείω μέρη ἔχει καὶ μὴ ἔστιν οἷον σωρὸς τὸ πᾶν [10] ἀλλ᾽ ἔστι τι τὸ ὅλον παρὰ τὰ (Aristotles' *Metaphysics* [*των μετὰ τὰ φυσικά*], Book VIII, Section 1045a, lines 8–10).

Aristotle's wholeness is not the same as Goethe's. In Aristotle's whole, a living thing needs more than just its organic parts to survive in order to make it "whole", that is alive.

20 The term "Mind's eye" is often used when referring to Goethe's method. One such author, Agnes Arber, links Goethe's usage of the Mind's eye to the Greek term Θεωρια (theory), which "was used both for scientific investigation and also for beholding". She understood this to mean that the Greeks "understood how to think with the mind's eye" (Arber 1950, p. 210). This shows that Goethe sees theory (Θεωρια) as a something we do, rather than in its present-day meaning, namely, a system of explanations (see below).

Goethe uses a variation of mind's eye, namely,

> We learn by seeing with our Mind's eye [Augen des Geistes], without which, everywhere, particularly in natural history, we blindly fumble [Wir lernen mit Augen des Geistes sehen, ohne die wir, wie überall, so besonders auch in der Naturforschung, blind umher tasten].
>
> (LA I 9, p. 138)

> It was instructive and satisfying to investigate the traces of the anatomical parts of mammals, how they are be discovered with our Mind's eye [Augen des Geistes], sketches of jaws recorded by memory alone, presented to us by our senes, which previously the imagination was unable to do [Es war belehrend und erfreulich nach den Spuren dieser Abteilungen auch bei Mammalien zu forschen und, wie man sie mit den Augen des Geistes zu entdecken glaubte, auf manche Kinnladen in- und auswendig aufzuzeichnen und so bestimmt dem Sinnen darzubringen, was vorher die Einbildungskraft zu bezeichnen und festzuhalten kaum in Stande war].
>
> (LA I 9, p. 178)

> The surgeon must know how to find the inner wound with the Mind's eye [Geistesaugen], sometimes without the help of [their] sense of touch; thus [they] find it necessary to apply the most exacting knowledge of detail to gain a kind of penetrating omniscience (trans. Miller in

Goethe 1995, p. 127) [Der Chirurg muß mit Geistesaugen, oft nicht einmal vom Tastsinn unterstützt, die innen verletzte Stellen zu finden wissen und sieht sich daher genötigt, durch strengste Kenntnis des Einzelnen sich eine Art von durchdringender Allwissenheit zu erwerben].

(LA I 9, p. 366)

Whether it is "Augen des Geistes" or "Geistesaugen", Goethe is referring to a form of practise that is equivalent to Anschauung and the Urphenomenon, namely an activity and not a supposition or system of explanations.

21 "Wenn ich mich beim Urphänomen zuletzt beruhige …" (LA I 11, p. 358). You could say the same with "walking" or "thinking", such as, "When I finally rest from walking".

22 "Die Glieder aller Geschöpfe sind so gebildet, daß sie ihres Daseins genießen, dasselbe erhalten und fortpflanzen können, und in diesem Sinn ist alles Lebendige vollkommen zu nennen" (LA I 10. p. 125; see also Goethe 1995, p. 22).

23 Ich möchte also wohl sagen: Schön nennen wir ein vollkommen organisiertes Wesen, wenn wir uns bei seinem Anblicke denken können, *daß ihm ein mannigfaltiger freier Gebrauch aller seiner Gleider möglich sei, sobald es wolle*, das höchste Gefühl der Schönheit ist daher mit dem Gefühl von Zutraun und Hoffnung verknüpft. (LA I 10, p. 127)

24 The concept of wholeness is thoroughly discussed by Bortoft (1996, 2012).

25 Goethe had used other terms interchangeably with the Urphenomenon, such as "pure phenomenon" [reine Phänomen] and "Archetype" [Typus].

26 Erster Entwurf einer allgemeinen Einleitung in die vergleichende Anatomie, ausgehend von der Osteologie (LA I 9, pp. 119–151).

27 Versuche zur Methode der Botanik: Ordnung des unternehmens (LA I 10, pp. 129–131).

28 Betrachtung über Morphologie überhaupt (LA I 10, pp. 140–144).

29 "… without [geology] the famous Morphology would never be complete." ["Es ist mir sehr lieb, daß ich so zufälligerweise diese Betrachtungen erneuert habe, ohne welche denn doch die berühmte Morphologie nicht vollständig werden würde."] Goethe to Schiller 12.11.1796 (WA IV 11, p. 260).

30 "… denn indem ich mit demselben Kennern der Botanik von meinen Bemühungen Rechenschaft geben wollte …" (LA I 3, p. 3; Translated by Herb Aach in Goethe 1971, p. 13).

REFERENCES

Arber A (1950) *The natural philosophy of plant form*. Cambridge University Press, Cambridge.

Arber A (1957) *The manifold and the one*. John Murray, London.

Bortoft H (1996) *The wholeness of nature*. Floris Books, Edinburgh.

Bortoft H (2012) *Taking appearance seriously*. Floris Books, Edinburgh.

Brady RH (2001) Getting rid of metaphysics. *Elemente der Naturwissenschaft*, 2: 61–78.

Burdach KF (1800) *Propädeutik zum Studium der gesamten Heilkunst*. Breitkopf und Hartel, Leipzig.

Cantor PA (2004) The scientist and the poet. *The New Atlantis*, 4: 75–85.

Ebach MC 2005 Anschauung and the archetype: The role of Goethe's delicate empiricism in comparative biology. *Janus Head*, 8: 254–270.

Fink KJ (1991) *Goethe's history of science*. Cambridge University Press, Cambridge.

Goethe JW (1971) *Goethe colour theory* (Ed. R Matthaei). Studio Vista Ltd, London.

Goethe JW (1995) *Scientific studies (Vol. 12). Goethe: The collected works* (Ed. D Miller). Princeton University Press, Princeton, New Jersey.

Haeckel E (1876) *The history of creation* (2nd edition). Henery S. King and Co., London.

Hart SP (2024) Goethe's platonic natural philosophy: How Goethean science provides an alternative conception of the cosmos. *Religions*, 15: 355. https://doi.org/10.3390/rel15030355
Lubosch W (1919) Was verdankt die vergleichend-anatomische Wissenschaft der Arbeiten Goethes? *Jahrbuch der Goethe Gesellschaft*, 6: 157–191.
Magnus R (1949) *Goethe as a scientist.* Henry Schuman, New York.
Oxford English Dictionary. Theoria (2025), September 2025. https://doi.org/10.1093/OED/4906011525
Sepper DL (1988) *Goethe Contra Newton: Polemics and the project for a new science of color.* Cambridge University Press, Cambridge.
Sepper DL (2009) Goethe, Newton, and the imagination of modern science. *Revue Internationale de Philosophie*, 63: 261–277.
Steigerwald J (2002) Goethe's morphology: Urphänomene and aesthetic appraisal. *Journal of the History of Biology*, 35: 291–328.
Trop G (2013) Poetry and morphology: Goethe's "Parabase" and the intensification of the morphological gaze. *Monatshefte*, 3: 389–406.
Wells GA (1978) *Goethe and the development of science 1750–1900.* Sijthoff and Noordhoff, Alpha can den Reign, The Netherlands.
Will F (1956) Goethes aesthetics: The work of art and the work of nature. *The Philosophical Quarterly*, 22: 53–65.

2 Goethe on Geology

What is the inference? Only this, that geology partakes of the uncertainty which pervades every other department of science.[1]

(LA I 2 p. 151)

In 1770 Goethe and his two friends, Johann Konrad Engelbach (1744–1802) and Friedrich Leopold Weyland (1750–1785) both native to Lower Alsace, set off to explore Alsace-Lorraine.

> We heard of the rich coal-pits at Dutweil [Dudweiler, Germany], of the iron and alum works, and even of a burning mountain; and we prepared ourselves to see these wonders close up.[2]

It is uncanny that a 21-year-old student studying jurisprudence at the University of Strasbourg had an interest in mining and metallurgy.

> In the alum-works we made accurate inquiries about the extraction and refining of these necessary materials, and when we saw great heaps of a white greasy, loose, earthy matter, and asked about its use.[3]
>
> … we entered into a chasm and found ourselves in the region of the Burning Mountain [Brennender Berg; a coal seam that has been burning since 1668]. A strong smell of sulphur surrounded us; one side of the cavity was almost red-hot, covered with reddish stone burnt white; thick fumes arose from the crevices, and through our heavy soles we felt the heat of the ground.[4]

The accounts by Goethe above were a distant memory when they were recalled 40 years later in his autobiography *Poetry and Truth* [*Dichtung und Wahrheit*]. It is unclear whether Goethe's reminiscing is a true account of his interest in geology, mineralogy and metallurgy early on in 1770, or whether he has in fact editorialised his own biography 40 years later (there are few letters that account for his interest in geology at the time). At the time of writing his autobiography (1811), Goethe had considered himself as much of a scientist as he was a writer of poetry, novels and plays. He had published the *Metamorphosis of Plants* (1790) and his *Theory of Colours* (1808–1810) and by the end of his life Goethe had amassed one of the largest collections of minerals and rocks in Europe (Hamm 2001).

Goethe never returned to the coal-pits of Dutweil. Armed with a doctorate in jurisprudence awarded in 1771, and a spell as an intern at the Imperial Supreme Court in Wetzlar (1772) as well as a trip to Switzerland (1775), Goethe arrived in Weimar on 7 November 1775 at the bequest of Carl August, the Grand Duke of Sachsen-Weimar-Eisenach. By 1779 Goethe was enlisted on to the Mines Commission in a legal advisory capacity (Boyle 1991), a time that Max Semper describes as Goethe's First period (1779–1789) of his geological studies (Semper 1914).

DOI: 10.1201/9781003275343-2

Goethe did not feature as a significant actor in the history of geology. Writing almost 70 years after his death, Karl Alfred von Zittel (1839–1904) wrote in his *Geschichte der Geologie und Palaeontologie bis Ende des 19 Jahrhunderts*:

> "As a student of Werner, Goethe was not comfortable with the teachings of the Plutonists, even if he saw through the unsustainable position of the gruff Neptunists" [and] "Goethe's achievements in the field of geognosy are of no significant importance".[5]

As is often the case with scientific histories, scientists and historians are concerned more with those who have influenced or changed a field, rather than document those who practised science at a particular point in time. While the former gives an entertaining account of how, for instance, certain people, such as Alfred Wegener, had rattled geologists, the latter investigates how people practised science and made sense of the natural world. Goethe did not make any significant contribution to geology or make any notable discoveries. Goethe's theoretical ideas about the formation of rocks were unduly influenced by the work of Abraham Gottlob Werner (1749–1817) and the Neptunists. Regardless, Goethe had an interesting way of observing the natural world, one that invites us to see rocks and minerals in an aesthetic way.

Goethe attempted to outline an autobiographical sketch of his geological studies. In Jena on 7 October 1820, Goethe penned *My relationship to science, and to geology in particular*.[6] The sketch reveals that Goethe had educated himself in his own way "without adopting any given or traditional approach". He had an "interest in natural objects and other things of the visible world" and "after mention my osteological illustrations, I should speak here of a similar project applied to the skeletal structure of the earth, geology". The truncated sketch continued to outline his geological studies chronologically:

> Mines at Ilmenau.
>
> Application to the study of the Earth's interior; how it reveals itself outwardly or has been explored within.
>
> First journey in the Harz mountains during winter; a dithyrambic poem about this journey still exists.
>
> Ongoing observation of rock forms.
>
> Masses which split apart.
>
> Conviction that this splitting follows certain laws.
>
> Difficultly in expressing this.
>
> Attempt to do so.
>
> Rock fissures which run vertically or nearly so.
>
> Oriented more or less clearly toward the cardinal points of the compass and traversed by other fissures (seldom at right angles, most frequently at an acute angle, thereby creating rhomboidal fragments).
>
> Many observations made to provide more positive information on the orientation of the above shapes toward the cardinal points of the compass.
>
> It seemed apparent that the alignment of the fissures toward the north occurred during the solidification of the rock (with the oblique east-west clefts not crossing at right angles however, thus creating the rhomboidal splintering).
>
> Proposal to create a model.
>
> Preliminary studies for this to be made from nature.

> We therefore needed to assemble accurate drawings.
> Trip to Harz mountains in August 1784 with Councillor Kraus.
> Short biographical sketch.
> Artistic, social qualities of this man.
> All these drawings were intended to show the details of fracturing, division, and form in groups of mountains and rocks. This was also the subject of diary notations (unfortunately all too brief).
> The diary to be published with notes clarifying the intention, and at the same time the drawings to be shown in a clear and useful way for future reference.
>
> *(Goethe 1989, pp. 138–140)*

Goethe's autobiographical sketch is a fair account of his geological studies, even though it leaves out several of his geological musings, such as his encounters with volcanoes in his *Italian Journey*. It would be prudent to flesh out Goethe's autobiographical sketch into a biography of his geological writings, including details that the polymath had wittingly or unwittingly left out.[7]

It would, however, be brazen to discuss Goethe's geological work without mentioning the books of German geologists Max Semper (*Die geologischen Studien Goethes*), Wolf Jürgen von Engelhardt (*Goethe im Gespräch mit der Erde*) and the extensive catalogue of Hans Prescher (*Goethes Sammlungen zur Mineralogie, Geologie und Paläontologie*).[8] Each attempted a biography of Goethe's geological writings. Semper (1914) separated Goethe's geological studies into two time periods, namely, the first (1779–1789), which included his studies in Ilmenau, Thuringia, the Harz and Italy; the intervening period (1790–1806), which included a third trip to Switzerland; and the second period (1806–1832), which saw a renewed and invigorated interest in geology,[9] including his geological system. To distinguish his own work from that of Semper, von Engelhardt attempted to write a chronology of Goethe's writings and his experiences of landscape and nature, in particular geography and geology.[10] Further summaries of Goethe's geological work, namely Magnus (1949, pp. 200–221), Wells (1978, pp. 47–74), Fink (1991, pp. 14–21) and Baldridge (1984), also place Goethe's ideas into context with the geology of his day.

Goethe had spoken of a method he had developed in his *Farbenlehre* or, *Theory of Colours*, one that embraced Anschauung and the idea of the Urphenomenon, and since his "interest in natural objects and other things of the visible world", had applied it to plant and animal morphology with great success (Goethe 1817). In fact, in 1795 he coined the term "Morphology" to mean the study of form.[11] Applying these ideas to the "skeletal structure of the earth" was challenging, simply because it required moving from the general to the specific, similar to Étienne Geoffroy Saint-Hilaire's notion of a unity of plan (i.e., an archetype or Urphenomenon). For example, different manifestations of the *same* form provide evidence for an Urphenomenon. The alternative, moving from the specific to the general, assumed that unity can be found in each individual organism or object. In the case of Georges Cuvier, who championed this view, a whole organism could be recreated based solely on an individual part (see Appel 1987). In modern-day geology the general lies in the specific. An individual rock may tell us about past environments, depositional and lithological processes, relative and, depending on the type of rock, absolute time. Without appearing to be Whiggish, Goethe's geological practise essentially failed to explain geological processes from a modern perspective. It did, however, force Goethe as a geologist to become far more observant and reject speculative hypotheses. In turn, Goethe did "get it right" in terms of framing minerals and rocks in a lithological sense. Whether Goethe did practise a geology that modern or 19th-century geologists approve of is not the point of this chapter. Rather, Goethe's method presents a unique perspective of the Earth or *geognosy*,[12] even if these observations are shoehorned into a Neptunist perspective.

Goethe's geological *opus magnus* consists of a theoretical, mostly Neptunist, perspective. The opus has various geological travelogues; lists of minerals and rocks; observations of rocks, minerals (mostly granite and basalt), as well as rock formations; reports from the various mines and quarries in Ilmenau and elsewhere; critiques of geological works; and an outline for a general geological text and various lectures.

MORPHOLOGY AND NEPTUNISM

Goethe reveals his own Neptunistic approach to geology in his short essay *On Granite* (1784),

> … anyone who knows the fascination natural mysteries hold for a man will understand why I have departed from my usual realm of observation and turned with passionate fervour to this one.
>
> *(Goethe 1995, p. 132)*

Goethe was a Neptunist, a follower of Abraham Gottlob Werner (1749–1817), who believed that granites were formed in a vast ancient ocean and that all rocks since derived from the weathered sediments of this primordial rock. Goethe's belief in Neptunism remained unshaken throughout his life, yet it exposed an Achilles heel in his own scientific approach.

Goethe was a pragmatist and an empiricist at heart. He craved the attention of the scientific community and was a member of several geological societies (see below). Goethe had one of the largest rock and mineral collections in Europe, which thankfully are still located in his home in the Frauenplan in Weimar. Yet, Goethe's approach to science was to expose how scientists see and appraise natural phenomena. Comparative anatomists of his day worked to understand the morphology, a term Goethe coined, of plants and animals. Goethe too contributed to this by discovering, along with Loader, the intermaxillary bone in humans. Where Goethe diverged from many of his contemporaries was in his pursuit of aesthetic appraisal, rather than in mechanising the natural objects.

How we observe and recall our sensory experiences of phenomena are at the heart of Goethe's scientific approach. Goethe wanted to explore what makes us scientists, how we see and recall what we have seen. Read any taxonomic or comparative biological text and you will notice that there is no methods section. Some texts may mention a microscope or a hand lens, but none mention the observers themselves. How we observe is hidden from our scientific audience. Taxonomists never openly talk about how they observe and recall, whereas art dealers, taxonomists of manmade objects, do. Art appraisers explore the textures left by the paintbrush in the paintings and the thumb prints left in the fired clay. Our ability to appraise the natural world is never discussed among scientists dare it taint their scientific discoveries.

Goethe wrote the missing methods section for comparative biology, colour and meteorology. That hidden scientific aesthetic appraisal is our ability to observe and recall what we observe in our mind's eye (see Steigerwald 2002). Goethe called this activity Anschauung. We use Anschauung to appreciate natural phenomena. For example, we can observe a seed sprouting in cotton wool, we can plant the sprouting seed and watch it grow into a plant. We can observe the plant growing leaves and flowers. We can see it wither and die. We can recall those changes and replay them in our mind's eye. Those individual moments constitute a whole, which is called the Urphenomenon. If we now attempt to picture the plant with the flower of another plant, we have created something we did not actually see. We have moved outside the perimeters of our experience, that is, we have moved beyond the Urphenomenon. What lies beyond the Urphenomenon is speculation and inference.

We need inference to make sense of what happened in the past (i.e., retrodiction) and what lies in the future (i.e., prediction). These are invaluable tools, but they lie outside our experience, and therefore, beyond the Urphenomenon. We often make the mistake of confusing a lived experience with an imagined one. Take the jointing of granite as an example. Over time the pressure on a granite dissipates and the rock starts to crack and form straight fissures. Water moves in and the rock weathers and erodes. The potassium feldspars hydrolyse and turn into kaolinite, which later forms clay. Granites on hillsides may form large mounds called *tors*. Over time these tors, made from blocks of granite, collapse. Boulders may roll down hillsides or collapse into voids left by other rocks long since gone. You can easily reconstruct a collapsed granitic tor as Goethe did through inferences. Goethe's reconstruction shows how the tor collapsed through time. But other than the present-day tor itself, the reconstruction lies beyond our lived experience and beyond the Urphenomenon. Unlike a seed, we cannot experience a granite forming from a melt. We cannot experience minerals crystallising as the melt cools. We cannot experience the granite moving through the bedrock. We cannot experience jointing and boulders breaking down. Time and physical pressure do not allow us to experience the life cycle, if you wish, of a granite.

Goethe had exposed scientific aesthetic appraisal in his study of morphology. Morphology was to be universal and the Urphenomenon was to present itself in plants, animals and in the inorganic. Here we come to Goethe's dilemma: where is the Urphenomenon in geology? Of the hundreds of pages Goethe wrote about geological processes, he was unable to find the Urphenomenon. Goethe made his concerns known in a letter to Friedrich Schiller (1759–1805) on 12 November 1796,

> … without [geology] the famous *Morphology* would never be complete.[13]

Goethe's "passionate fervour" of Neptunism let him down, as it got in the way of Anschauung and jeopardised his universal morphology. Goethe became frustrated in his attempt to reconstruct the history of granites and other rocks and minerals. What he discovered in his studies on colour, plants and animals was absent in geology. Yet he persisted from 1770 until his death in 1832. Poring through his writings and dare

I say, on occasions, scribblings, Goethe did undertake Anschauung. I am not certain whether Goethe was aware of it, but he did observe many minerals and rocks in Europe's most prized collections. Goethe made a plethora of lists and notes of what he saw. These included travelogues describing the rocks and minerals he saw in the field.

Anschauung is essential for the scientific appraisal of minerals and rocks, in the same way it is today. Think of the way a geologist identifies a mineral. If the mineral baffles them, they instantly test for specific gravity (density) by holding the mineral, they scratch the surface to check for hardness, hold it up to the light to check for lustre and diaphaneity (transparency) or determine its colour and crystal shape. They may scratch the mineral with an unsealed white tile to test for streak colour or use a magnet to see if it is magnetic. Some minerals give off an odour, others you can taste. How can you tell a siltstone from a mudstone? Take a piece from each and chew. The siltstone has grit in it, the mudstone is, well, muddy. There is no evidence to show whether Goethe did any of this or whether he knew about such tests.

Goethe's geology is often dismissed by historians of science as having contributed little to the field.[14] I think the opposite can be argued. While Goethe made no scientific discoveries of a geological nature, he did discover that there are limitations in geology, particularly when it comes to morphology. We cannot observe geological processes, yet we can infer them, but we must not confuse observation with inference. Goethe made this mistake in his geology by his stark adherence to Neptunism. The Urphenomenon is where it always was, in our Anschauung of rocks and minerals, that leads to their discovery, identification and eventual classification. Goethe knew about the different types of schists or hornblendes. He could identify and classify minerals and rocks using Werner's system (Werner 1787), before sophisticated mineral classification systems, such as that by James Dwight Dana (1813–1895), which appeared long after Goethe's death.

Morphology and its tools, Anschauung and the Urphenomenon, are all that is needed to observe, measure, record and recall the natural world. Shoe-horning Werner's theory of Neptunism into Morphology severely compromises its integrity. Unfortunately, Goethe never was able to rise above the confusion.

> [M]ineralogy has had an interest for me in two respects: first, because of its great practical application, and because therein can be found a document concerning the formation of the primeval world. … Now, however, since the death of a remarkable man [Werner], everything in science is turned upside down; I will not continue publicly in this subject, but rather cease, to be at peace in my conviction.
>
> *(LA I 2, pp. 357–580, trans. in Baldridge 1984, p. 167)*

One particular idea that prevailed within Neptunism was the idea of a primordial or primary rock (Urgebirge), namely, the first to crystallise out of the ancient sea. These primary rocks are considered the source material for other rocks, as Goethe notes:

> Granite was created through crystallisation […]
> When finally the heavy flotsam from above settles on the granite, only a trace of the original crystallisation remains.
> The entire structure of our Earth can be explained by crystallisation.[15]

In his *Brief classification and description of the different types of rocks*,[16] Werner's classification recognised four main divisions of rock types or Gebirgsarten. "These rock types are primary, secondary, volcanic and alluvial rocks".[17] Primary rocks (uranfänglichen Gebirge) consisted of "granite, gneiss, mica-slate [schist?], phyllite, porphyry slate, porphyry basalt ..." (Werner 1787, p. 6). Secondary rocks (Flözgebirge) included sandstone and other types of layered rocks (strata). Volcanic rocks were followed by the "very last, namely, the alluvial rocks [aufgeschwemmten Gebirge], that we can assume as the newest" (Werner 1787, p. 26). But there was a problem with how these rocks were ordered.

> The primary, secondary, and alluvial rock types merge into one another, and one has primary rock types that can almost justifiably be called secondary rock types; and then again, one becomes doubtful whether one should count secondary rock types among the alluvial rocks. Some of the first, namely the primary rock types, also gradually change into secondary rock types. According to the way in which these types of rocks were formed, which changed from one to the other almost imperceptibly during the enormous period of our Earth's existence, it is also possible that such *transitions* have to take place in these types of rock.[18]

Later classifications, such as Brochant (1803) and Ludwig (1804), cobbled together from various other Wernerian works,[19] introduce a new second class or division, namely, Übergangsgebirge or transition rocks (see also Jameson 1808), which formed between the primary and secondary divisions. Transition rocks included limestone, greywacke[20] and transitional trap rocks, that is, microcrystalline intrusive igneous rocks, such as dykes and sills.

Goethe found the new division intriguing,

> These so-called transitional rocks [Übergangsgebirge] are extremely strange to me, because I am only now getting to know its peculiarities, having previously only dwelled on primary and secondary [rocks].[21]

Interpreting geological formations becomes impossible if the lithosphere is seen as a single continuous temporal formation, in which granites were the first rocks to form, then followed by transitional, secondary and alluvial rocks (Table 2.1). Werner imposed time onto rocks, meaning that all granites, no matter their position in a stratigraphic column, are *by definition* the oldest, even if it makes no logical sense. If, however, we were to use the superposition of rocks as characteristics of time, then it would be much easier to interpret the sequence of rock formation. Ironically, Goethe used the latter to interpret the weathering, erosion and transportation of granite tors, yet he didn't apply the same logic to stratigraphy.

Goethe's dedication to Werner's impractical Neptunism, particularly the granite as the primary rock, made it difficult for the German polymath to advance his morphology in geology. Goethe fell into a Wernerian slumber, but would he ever awake?

With Goethe lost in slumber, the language surrounding Goethe's scientific writing, in particular the prefix *Ur-*, meaning "the first" or "earliest", has been a distraction for many readers of his work. As I will show throughout this book, the prefix *Ur-* highlights some of Goethe's most poorly thought-out ideas that were detrimental

TABLE 2.1
The Wernerian System (1787) Using the Terminology from Bakewell (1815)

Werner's (1787) Classification System

Uranfänglichen Gerbirgsarten [Primary Rocks]	Flözgebirgensarten [Secondary Rocks]	Vulcanische Gerbirgsarten [Volcanic Rocks]	Von dem aufgeschwemmten Gerbirgsarten [Alluvial Rocks]
1. Granit [1. Granite]	1. Flöz-Kalkart [1. Calcareous Limestone]	1. Aechtvulkanische [1. True Volcanic]	1. Seiffengebirgsarten [1. Diluvium/ Placer Deposit]
2. Gneis [2. Gneiss]	2. Sandsteinarten A. gemeine Sandstein [A. Silicious Sandstone] B. Grau-wakke [B. Greywacke] C. Puddingstein [C. Conglomerate[a]/ Breccia]	2. Pseudovulkanische [2. Pseudovolcanic]	2. Gerbirgsarten des niedrigen Landes A. Sandland [A. Sand] B. Laimland [B. Loam[a]] C. Turf-Erde [C. Peat or Wood-coal]
3. Glimmerschiefer [3. Micaceous Schist]	3. Steinkohlen-Flötzarten [3. Coal]		
4. Thonschiefer [4. Slate]	4. Kreide-Flötzarten [4. Chalk rocks]		
5. Porphyrschiefer [5. Argillaceous Schist]	5. Steinsalz-Flötzarten [5. Rock Salt]		
6. Porphyr [6. Porphyry]	6. Gips-Flötzarten [6. Gypsum]		
7. Basalt [7. Basalt]	7. Eisenthon-Flötzarten [7. Ironstone]		
8. Mandelstein [8. Amygdaloidal rock]			
9. Serpentinstein [9. Serpentine]			
10. Uranfänglicher Kalkstein [10. Primitive limestone]			
11. Quartz [11. Quartz]			
12. Topazfels [12. Topaz rock]			

[a] Denotes terms not used by Bakewell (1815).

to his *Morphology*, and continues to be misinterpreted by authors as a proto-evolutionary concept (i.e., Urpflanze), one that foreshadows Darwinism, evolutionary processes (e.g., Haeckel's Biogenetic law) or evo-devo[22] (e.g., Haeckel 1876, p. 82; Niklas and Kutschera 2017).

URGESTEIN

The term Urgestein[23] is often attributed to Goethe, for example,

> Goethe saw in granite the Urgestein[24]
>
> [Goethe] tried to intervene in the dispute with his own theses on granite and claimed that granite is a Urgestein ....[25]

Yet, Goethe rarely used the term Urgestein,

> ... that through partial dissolution as well as partial persistence of the Urgestein[26]
>
> We took samples of all the Urgestein with us and left a box in Eger.[27]
>
> *(Goethe to August Goethe, 29 April 1820)*

> Furthermore, last year, Mr. Count Sternberg drew the attention of nature lovers who were engaged in geognostics in Marienbad to the slow but great power that the gases exert on the Urgestein.[28]

In his extensive mineral and rock collection, Goethe only noted three specimens as *Urgestein*, namely, "84. Contiguous [?] Urgestein" collected from Marienbad [Mariánské Lázně, Karlovy Vary Region, Czechia] in 1822, "5. Urgestein altered beyond all recognition with a fresh fracture", and "6. The same in round ball", both from Altalbenreuth [near Mýtina, Cheb Region, Czechia].[29] Given the popularity of "Goethe's Urgestein", the polymath himself rarely used the term in his geological investigations.

Why has the term Urgestein become so ubiquitous in the literature on Goethe's works, and what did Goethe mean by the term? To understand the relevance of Urgebirge, we need to understand where it originates.

The term Urgestein may be traced to the term "primitive rock" in English by Robert Bakewell in his *An Introduction to Geology* (Bakewell 1815). In his seminal text, Bakewell describes primitive rock as one of the three classes of rocks. "The classes are denominated primary or primitive rocks; transition or lower secondary rocks; secondary stratified rocks, called by the Germans Flœtz rocks; volcanic rocks, and alluvial ground" (Bakewell 1815, p. 7). "In fact, Sienite [sic] and porphyry are called by Werner with his primitive rocks" (Bakewell 1815, p. 63), as well as "1. Granite. 2. Gneiss or slaty granite. 3. Micaceous schist or mica slate" (Bakewell 1815, p. 61). Bakewell is clearly deriving his terms from Werner.[30] Yet, when we read Werner we see that primitive rocks, transition rocks and secondary stratified rocks are termed uranfänglichen Gebirgsarten (Werner 1787, p. 5) or Urgebirge, Übergangsgebirge and Flötz, respectively (Leonard et al. 1817, p. 160). The term Urgestein, rather than its definition, possibly has its origins elsewhere.

Meixner (2002) incorrectly cites *Granit I* (LA 11A, pp. 9–10) and *Granit II* (LA I 11, pp. 10–14) as the source of Goethe's Urgestein: "we find similar composites above all in the geological writings, where granite is described as the Urgestein" (Meixner 2002, footnote 9). The term is not used in either entry. What Meixner (2002) does infer is that Urgestein is most likely influenced by Goethe's other ubiquitous terms, namely the Urphänomen and Urpflanze, and less so, the Urtier or primordial animal.[31] If this is the case, the Urgestein would be an older term than Urphänomen and Urpflanze. Goethe first used Urphänomen in his *Theory of Colours* or *Farbenlehre* in 1810 (LA I 4, p. 71), Urpflanze in his *Italian Journey* or *Italienische Reise* in 1816–1817 (Urtier first appeared in *Zur Morphologie* 1817). Nonetheless, the term Urgestein is found in earlier geological texts such as Schrank (1793) and Reimarus (1802), suggesting that the term was already in use within the geological literature, but not used by Goethe until the 1830s.

The old chestnut, paraphrased as "Goethe saw granite as *the* Urgestein", may be both redundant and obvious. If by Urgestein we mean Urgebirge,[32] then yes, all followers of Werner's theory would consider granite, as well as gneiss and schist, as types of Urgestein. Granite, however, resonated over these other types of primitive rocks. In *On Granite* (*Über den Granit*, 1784), Goethe called granite the "basis of all geological formation".[33] A more poetic account is given in *Return to Granite* (*Rückkeher zum Granit*, 1807), where we see the importance of granite to Goethe:

> Granite is a diverse rock.
> So common and rarely shown.
> Accurately observed. Enter hornblende: syenite.
> Or it becomes slate, turns into gneiss.
> Or contains its parts larger, magnified by itself, called porphyry granite.
> So that one really has to pay attention to the instantaneous becoming, the fluctuation, the transition. Samples of actual granite.[34]

Also here,

> This granite is also found at its transition into gneiss. The twinning,[35] in their still completely decided form and properties, are influenced by the mica, whose flat direction begins to manifest itself here. They themselves appear intertwined, stretched because the mica has passed into them and acted on them. They, on the other hand, determine the shape of the whole rock, in that the flaky part of it is produced entirely by them.[36]

Goethe's mineralogy is spot on if, and only if, we interpret "geht in Gneis über" ["turns into gneiss"] and "das Übergehen" ["the transition"] to mean Steigerung in the Goethean sense. The addition of hornblende means granite grades into the more mafic syenite, an intermediate igneous plutonic rock. The idea that the larger potassium feldspar crystals in porphyry granite are endemic (i.e., not xenocrysts), and had crystallised before mica, all fits into a 21st-century understanding of petrology.[37] Even the notion that slate, or even granite, turns into gneiss fits into a regional metamorphic facies model. The notion that the Urgestein or Urgebirge starts off with granite and transforms into other rocks does not necessarily equate to Goethe's morphology as outlined in his 1817 opus *Zur Morphologie*. In order to distinguish

Goethe's morphology from contemporary ideas at the time we need to investigate his ideas on granite.

ON GRANITE

Goethe's *Granit I* and *Granit II*,[38] written in 1784, were composed before his morphological ideas on the metamorphosis of plants in 1790, the Urphenomenon in 1810 and the archetype Typus in 1817 were developed. We see these ideas explored rather momentarily in *Return to Granite*, and rarely after 1807.[39] Goethe's morphology of the general type, metamorphosis and archetype doesn't translate well into geology, even if an unobserved transition, such as granite to syenite, can be retrodictively assumed. Here we come to Hartmut Böhme's thesis, namely,

> Goethe fulfils the temporalisation of natural history. However, he does not do it primarily to expand empirical knowledge of the origin of the Earth [...] he tries to connect these cognitions, which are unimaginable in sensory terms and therefore sublime in the Kantian sense, with the tension between the immemorial ground of being [Seinsgrund] and pathetic existence, which can certainly be experienced. To do this, he does not use granite as a scientific object, but as a symbolic form, not in the form of conventions or laws [*thesei* or *nomos*] but rendered from nature [*physei* or essence], and in this way makes it aesthetically evident. In this dialectic tension, Goethe reflects the configuration of history and nature, of temporality and the ground of being, of the subjective and the objective.[40]
>
> *(Böhme 2016, p. 57)*

What Böhme is referring to here is the *physei–thesei* or *physei–nomos* controversy, which harks back to Plato's book *Cratylus*. In it, Plato argues for two different approaches to etymology, or names in particular. The book opens with Hermogenes inviting Socrates to a discussion she was having with Cratylus. She explains that Cratylus argues that names are natural and not conventional. As the conversation between Hermogenes, Cratylus and Socrates unfolds, we discover two ways of viewing language: either as a natural phenomenon (i.e., *physis*), or something that is learned by ways of rules and conventions (i.e., *thesei* or *noma*).[41] The problem an observer, such as Goethe, has with observation is that it occurs in real time. We may observe a flower open, a fern frond unfurl or the wing-beat of a bird. We are unable to observe history in the sense of sediment lithifying into rock, or granite turning into a gneiss. The observer can *imagine* a historical event, but is unable to observe it directly, hence, "unimaginable in sensory terms". Böhme correctly identifies time (or history) as the issue within Goethe's morphology – it needs to be "seen" as relative, rather than as absolute.[42] How then does Goethe *see* the "temporalisation of natural history"?

Let's return to granite, how it forms and how it weathers and erodes. Each process takes a seemingly inordinate amount of time: millions of years for granite to cool, differentiate and crystallise, and for uplift of the solid mass to fracture and joint, and for its minerals to break down into clay and other sediment. Yet, we can factor time into our observations, particularly in the last of these processes. If we were to compare granite, for argument's sake, to language, we can get a sense of how Goethe may have interpreted history. We may observe granite, as Goethe did in *Granit I* – its minerals, their shape and composition.

> This remarkable kind of rock differs from all the others in that, while it is not simple, but consists of visible parts [minerals], the first sight shows that these parts are connected by no external means, but only exist side by side, and hold themselves together. We call these parts, which can be easily distinguished from one another: quartz, feldspar, mica, to which some are sometimes added as schorl [tourmaline].
>
> If we look closely at these parts, it seems to us that they were not in front of the whole, as one would normally think of parts; they do not appear to have been put together or brought together but rather came into being together with their whole to make it up. And although only the mica often appears in its six-sided, tabular crystallisation, and quartz, and felspar, lacked space in order to grow into their own, it is evident that the granite was created by a living crystallisation, which at its origin was very crowded. We have been able to draw some conclusions about its origin and the matter from which it arose.[43]

Goethe refers to mica, a group of silicate minerals. Given that he can see "its six-sided, tabular crystallisation" reveals that it is likely biotite, which is the first of these minerals to crystallise from the melt. The fact that "quartz, and felspar, lacked space in order to grow into their own forms" suggests that the feldspar is a potassium feldspar, possibly orthoclase, and crystallised after the biotite along with quartz. Yet, Goethe doesn't specify "its origin and the matter from which it arose".

The order of mineral growth, such as the order or sequence of crystallisation from a melt, or the principle of inclusions, hadn't occurred to Goethe even though he observed it in detail. For example, in *On the expression of different types of porphyry* [*Über den Ausdruck Porphyrartig*], Goethe discusses several more of these sequences.

> It would now hardly be audacious to call that mica slate porphyry, which is given a fibrous texture by quartz, because this quartz, although without a definite shape, has nevertheless emerged from the mica mass in plates and layers, just as one can say the opposite just as well because mica is contained in stronger quartz parts.[44]

We see a similar sequence of crystallisation in a previous passage by Goethe.

> The twinning, in their still completely decided form and properties, are influenced by the mica, whose flat direction begins to manifest itself here. They themselves appear intertwined, stretched because the mica has passed into them and acted on them. They, on the other hand, determine the shape of the whole rock, in that the flaky part of it is produced entirely by them.
>
> *(LA I 1, p. 383)*

Mica as a mineral group crystallises from melts before quartz, meaning that for the mica to be "contained in stronger quartz parts", it had to have formed first, followed by the quartz. In the second passage, Goethe notes that the mica forms after the twinning feldspar "because the mica has passed into them and acted on them". Mica is flaky and once foliated (the minerals along a single direction form a sheet of aligned crystals), as it is in gneiss, it forms a flaky feature as Goethe notes above. Reading Goethe, one may assume that he means that crystallisation occurs in a sequence over time, namely, feldspar (most likely the potassium-rich orthoclase), followed by muscovite mica, which is followed by quartz. A similar sequence was mentioned earlier by Goethe in his *Theory of Rock Stratification 1785*,[45] in which granites were followed by gneiss; and mudstones, siltstones and shales followed slate and phyllite.[46]

Goethe does subscribe to a logical ordering within Nature, one that uses relative time rather than absolute time.

> At the first crossing of the granite, i.e. the first appearing diverse, crystallisation occurs, i.e. the first successful individualisation of the Naur. In this last one, organic formations of both large ferns and corals can be seen.[47]

This logical ordering prevails throughout Goethe's *Der Dynamismus in der Geologie* [*The Dynamic in Geology*], and in it we can see several observations and reconstructions made by Goethe.

> Everything geological lies between an oldest and youngest; between the granite, which we know as the first, and last the diluvial mountains.
>
> *
>
> The main difficulty of geology lies in the strata; on the fact that the atomistic and mechanical, which of course proves to be effective in certain moments, is pushed back as long as possible, while the dynamic, a lawful-conditioned emergence, a development and transformation, is given its right.
>
> *
>
> If, through the atomistic view, one sees something that has already become drifting, depositing and solidifying, then the dynamic view, on the other hand, introduces the living play of the elements and their attractions into the moment of emergence. In it, a great deal can still be derived from the calm execution of internal laws, which in the former can only be made understandable through the effort of many floods and external forces.
>
> *
>
> Likewise the dynamic view teaches us to explain partial origination without difficulty, which is scarcely conceivable in the mechanical and tidal systems; namely, it considers all matter to be capable of life and transformation, depending on the conditions that bring about it; it does not deny the cosmic. It creates a play of elements throughout the entire atmosphere, with attractive forces towards the solid, as we become aware of them every day, only modified; she sees that there is an interaction between what is at hand and what is being created, through which the former can act on this, and the latter on that. Finally, in what has already been formed, it allows for an inner formation, that is, a collection and attraction of what is similar and corresponding.
>
> *
>
> The lowest, underlying thing that we found on Earth is the granitic. Its distinguishing concept is that it has no *continens* and *contentum*, but rather a complete being-in-one-another, a perfect trinity of its parts. They are equal in it and no one has a decisive preponderance over the other.
>
> *
>
> If the granite gives up this character, it happens because one of its parts gains predominance over the others, makes its way of being the dominant one and forces the others to shape themselves in this way. The granite, where it gives up its character, therefore has not one but multiple types of transformation.[48]

*

As the granite reveals [Dieses Aufgeben seines Charakters], diversity occurs more or less continuously. He has ceased to rule; now there is anarchy in which everyone strives for dominance. At this moment the metal formation appears.

*

This abandonment of one's character in granite, this metamorphosis, can be seen as a step out, a transcendence. To describe its diversity, think of it in the image of a sphere, which, as soon as you step out of its centre, admits radii to all ends.

*

If we see in it that the elements that produce granite can, under certain circumstances, be determined into something other than itself, then a backward determination, a redetermination of the other into granite, is also not completely unthinkable.

*

The first transgression [Überschreiten] of the granite occurs in the state in which mainly metal formation occurs. Although clay and pebbles predominate in it, lime, which has already been acidified in many ways, also occurs in it.

*

From this transgression a second thing happens. In this one there seems to be a more fluctuating formation than in the first. What has just been fixed is canceled out, consumed, disturbed and recreated by new causes. People usually try to explain it mechanically, but it still belongs to the dynamic. It is not a tearing and attachment from outside, but rather an internal dissolution and rebinding.

*

However, phenomena occur with and in it in which the friction and flushing of the floods cannot be denied, just more rarely than one would think, and many things that at first glance seem to be caused by floods are, on closer examination, better derived from weathering.

*

When the granite is first transgressed, the first appearance of diversity, namely crystallisation, occurs, that is the first successful individualisation of natural forms. The last formations are organic, consisting of both large bracken and corals.

*

The greywacke, a name for many formations, occurs close to the granite. She carries great metals within her.

*

The porphyry is also close to the granite and has all the same components of granite. In it one clearly distinguishes the *continens* from the *contento*.[49]

There are several concepts to unpack from Goethe's *Der Dynamismus in der Geologie*. Firstly, there is the term "dynamism". When Goethe describes what is dynamic, he is referring to past processes that are unobservable. Rather, he means the results of past processes, showing that granites were part of a dynamic interaction of parts that conformed to a natural law, that is, "a lawful-conditioned emergence, a development and transformation, is given its right". These terms, "emergence" and "transformation", are all made in hindsight as their dynamism is inferred rather than observed. Not observing the phenomenon from which granites emerged denies the observer Anschauung and the Urphenomenon, without which Goethe science of Morphology is null and void. Possibly to preserve some essence of Morphology, Goethe treats the granite as the fully formed plant, with mature leaves, a flower with sepals and attempts to reverse the process of metamorphosis as though it is an act of observation. When the granite reveals its characteristics, "it happens because one of its parts gains predominance over the others, makes its way of being the dominant one and forces the others to shape themselves in this way". We can see this in granites when older minerals, which have formed first, are apparently surrounded by other minerals that seemingly have formed around them. These interlocking crystals are characteristic of all intrusive igneous rocks and give us a glimpse into how the rock may have formed. In Figure 2.1 we can see that the pink orthoclase has formed individual crystals, in which the crystal habits are clearly seen. These crystal shapes can emerge from rock, in the same way that an unseen moth may emerge from the bark of a tree. The other minerals emerge and seem to form around the orthoclase, the black biotite and white plagioclase forming next, still able, in places to form crystals. Last of all, the clear quartz simply fills in the empty spaces and was unable to form any crystals due to the lack of space. As with the fully formed plant, we may play out the "metamorphosis" within the Urphenomenon. Granite, however, is different. There is no observed metamorphosis as seen in plants. Only the fully formed granite was ever observed, even though it may reveal the sequence of crystallisation events.

We return to *Granit I*, following "[w]e have been able to draw some conclusions about its [granite] origin and the matter from which it arose".

> Since the human being only sees such effects as result from great movements and violence of the forces, he is always inclined to believe that Nature uses violent means to produce great things, even though he could teach himself something else every day. Thus the poets presented to us a quarrelling, disunited, raging chaos.[50]

FIGURE 2.1 The interlocking crystals within a granite. The orange crystals are orthoclase, the clear greyish crystals quartz and the dark crystals biotite.

(Photo by the author, 2026.)

Are "… see such effects …" ["solche Wirkungen in die Augen fallen"] that Böhme calls "unimaginable in sensory terms"? Can the "… tension between the immemorial ground of being [Seinsgrund] and pathetic existence" actually be experienced? Goethe continues,

> My spirit has no wings to soar forth into those primordial beginnings. I stand firmly on [my theory of the formation of] granite and ask it if it will give us some reason to think about the nature of the mass from which it was made.[51]

There is certainly a lot to think about. Goethe understands that he has moved beyond phenomenon, with "no wings to soar forth into those primordial beginnings", yet he maintains his belief in the formation of granite based on a set of criteria. Here we return to Plato's *physis* and *thesei*, in the form of what can be seen and measured (*physis*) and convention (*thesei*). The struggle between *physis* and *thesei* may well be Böhme's "immemorial ground of being [Seinsgrund] and pathetic existence", respectively. Goethe rarely encounters this tension again in his latter work on colours, plants and animals.

If we return to Böhme's thesis, we find Goethe further fulfilling "the temporalisation of natural history" in his study of granite tors.

ON GRANITE TORS

Goethe, it seemed, had a fondness for granite tors (Figure 2.2). It featured prominently in his *Jointing of Granite* [*Zerklüftung des Granits*]:

Rock face
  its sides
    Edges.
Main joints[52]
Main horizontal joints
Main vertical joints

Connecting joints
Connecting horizontal joints
Connecting vertical joints

Massive rock[53]
  Main massive rock
  Intermediate massive rock

Cracks
Crevices
Gashes

Banks[54]
Plates[55]

Columns
Blocks
? Ridge
Knags
Ridge, summit
truncated.[56]

Following the poem Goethe describes a granite tor [Granitgebirge] in which he introduces terms for several of the structures. The term "wall" [Wände] is used to describe the rock face.

> The largest divisions of granite, to which we first direct our observation, run in the direction of the mountains and separate the rock from above into massive rocks, [Massen] which are narrow in length and height and whose sides run parallel to one another. We call these rock faces [Wände].[57]

Goethe continues to describe the jointing.

> The first divisions occur through horizontal joints, which are criss-crossed by vertical ones, create large parallelepiped [cubic] structures, or when the joints are oblique, form rhomboid blocks.[58]

FIGURE 2.2 Kew-Y-Ahn, a granitic tor in the historical village of Hartley, NSW, Australia. **(Photo by the author, 2023.)**

Goethe then introduces several new terms.

> The granite tors [Granitgebirge] do not consist of entire undivided massive rock [*Massen*], rather those are split in a variety of ways.
>
> The first main divisions are the same parts down the length of the tor. This creates walls of considerable width. The middle wall is mostly vertical, the following walls are partly vertical, while some of them tend to move in the direction of the tor. These walls are again criss-crossed by fractures. At first we notice some [fractures][59] move horizontally or roughly along a horizontal line, which I named horizontal joints [Flözklüfte] according to the common mining expression. Others [fractures] that run more or less vertically to the previous joints and intersect them can be called vertical joints [Gangklüfte]. This creates divisions that form large parallelepipedic [cubic] or rhombic massive rock, and these are separated again by further sub-divisions. This is why I should be allowed to name those first aforementioned joints as Main bedding joints [Hauptflözklüfte] and Main vertical joints [Hauptgangklüfte] respectively, and the massive rock [Massen] as Main massive rock [Hauptmassen].[60]

The rest of the terms used in the poem are described below.

> The subdivisions are not as simple or as regular and cannot be understood without the presence of the tor, nor in the absence of a model or a drawing. Soon the joints [Klüfte[61]] run in parallel with the main joints [Hauptklüften] within the connecting massive rock [Zwischenmassen], and slowly peter off. The joints divide horizontally into banks [Bänke] or perpendicular into even blocks [Blöcke], they then divide into several lengths, in which the joint running underneath is connected to a block that is, as it were, proposed as a ridge [Riegel]. Soon a crack [Riß] will run into the next massive rock, and soon it disappears. The first plate [Tafel] displays three such walls [Wände] that stop at the summit of the tor, with all its parts described in the above terms.[62]

Goethe had created a nomenclature for granite tors, or any rock for that matter, with which to accurately describe the morphology of rocks. Goethe considered the jointing to be the result of crystallisation, through "cracks and splitting through crystallisation, not through cooling"[63] (LA I 1, p. 96; also see Wells 1978, p. 73, footnote 7).

Goethe drew diagrams explaining the jointing in granite tors over time (Figure 2.3).[64] In his most descriptive drawing, Goethe shows how each block is eroded over time leaving behind the fallen mass of the tor.

Goethe's model is accurate, even by the standards of 21st-century geology. The process, however, is far more complicated. As the granite pluton cools, it contracts and forms vertical cracks. Hotter hydrothermal fluid from the mantle moves through the cracks, altering the rock on either side and leaving secondary minerals, such as quartz veins. As the granite is gradually uplifted or "unroofed", further pressure is released as the granite expands upwards, creating horizontal cracks. Groundwater moves along these cracks creating larger fissures, and minerals such as potassium feldspars turn into clay minerals that slowly erode from the granite, creating smaller and smaller blocks. Given these processes take millions of years, and are mostly underground, it isn't hard to imagine a large body forming megacrystals from smaller crystals.[65]

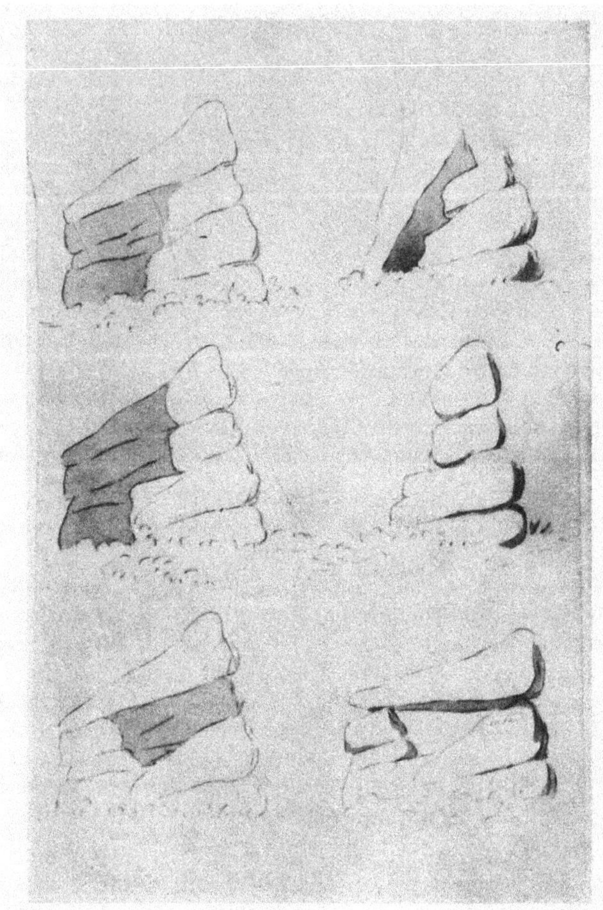

FIGURE 2.3 The origin of the granitic labyrinth [tors] on the Luisenburg in the Fichtel Mountains. Left: shaded areas on the left represent the unweathered blocks. Right: the weathered blocks that have in part collapsed.

(Drawn by Goethe on 4th May, 1820 (LA I 2, plate IV; Corpus Vb, p. 214a).)

Goethe attempted to apply his morphological method to granite, as he did to other rock types. We gain some insight into Goethe's later thinking via *Gestaltung unorganischer körper* [*Formation of inorganic bodies*] in 1820. Goethe never shifted away from thinking that granite, as well as sandstone, are products of inorganic crystallisation. As with the plant being "all leaf", Goethe saw rocks as being "all crystal". If soft clay breaks into even pieces as it dries, then all types of mountains, whether they are granite or sandstone, break into rhomboid blocks. Even mud bricks fired in ovens form columns. The separation of mass into parts, Goethe describes.

Formation of the mass is in itself crystalline.
Crystallisation in freedom.
Isolated crystallisation within the mass.

Porphyry types.
Tendency for traps [basalt] to form columns.
Tendency for every simple mountain type to form regular shapes.
Do not always appear.[66]

(LA I 2 pp. 175–176)

Goethe's morphological thinking stayed with him until his death and beyond. So too did his belief in Neptunism, which did impact his art, particularly his most famous work *Faust*. In Part 2, Act II of *Faust II*, *Classical Walpurgis Night*, a dream-like creation myth appears, which alludes to jointing.

Glad of any new terrain:
Give us cracks in the rock face [Felsenritze],
And the dwarves pop up again,
Male and Female, dwarfish twosome,
Toiling gladly cleft by cleft …

(Faust II, verses 7611–7615, translation by Luke 1994)

Later that night, in a dialogue between the pre-Socratic Greek philosophers Thales and Anaxagoras, it alludes to the Neptunist–Plutonist debate.[67]

ANAXAGORAS [*to* THALES].	I see you're of the same opinion still: What more's required to bend your stubborn will?
THALES.	The wave will bend at every wind's insistence; From the unyielding rock they keep their distance.
ANAXAGORAS.	I say this rock by fire was created.
THALES.	In moisture all that lives originated.
THE HOMUNCULUS.	[*appearing between them*]. Allow me to accompany your debate! I too am trying to originate.
ANAXAGORAS.	In one night, Thales, you I think would fail To make from slime a mountain on this scale.
THALES.	The peaceful flow of Nature's living powers Needs no constraint of nights or days or hours. She moulds and rules all forms, and even on The greatest scale no violence is done.
ANAXAGORAS.	It was done here! Monstrous Plutonian heat, Aeolian explosive gas, replete With rage, burst through the earth's old flat crust and so At once compelled this great new hill to grow.
THALES.	Well, so you say; what then? The mountain came, And when all's said and done, that no great shame. We waste time thus disputing, and mislead The patient flock who pay our words some heed.

*(Faust II, verses 7851–7872,
translation by Luke 1994)*

Thales of Miletus (~626–545 BCE), the elder and founding Greek philosopher, represents the Neptunist Werner, whereas Anaxagoras (~500–428 BCE) may be a parody of the Plutonist James Hutton. Thales is known for his break with mythology to explain natural phenomena, choosing instead to use deductive reasoning. Diogenes Laertius, in his *Lives of the Eminent Philosophers*, notes that "[h]e held that the original substance of all things is water …" (Diogenes Laertius, Book 1:27). Anaxagoras claimed "that the sun is a red-hot mass of iron and is larger than the Peloponnese", and that meteorites, such as the stone of Aegospotami, fell from the sun. His claim led him to be tried for impiety and banished from Athens (Diogenes Laertius, Book 2:8, 12–15).[68] Goethe used a similar motif in his eulogy to Werner in *Zahme Xenien*:

> As soon as the noble Werner turns his back,
> They destroyed the Poseidonian Empire;
> When all bow before Hephaestus,
> I can't do it at once ….[69]

Goethe's geology faltered in part because he failed to keep theory (i.e., Neptunism), separate from his method (i.e., morphology). While his work on granite did uncover a nomenclature for studying granite and models that retrodict past positions of boulders in tors, his ideas on basalt were driven purely through the lens of Neptunism.

ON BASALT

Goethe never fully woke from his Wernerian slumber, even when faced with empirical evidence. Although he disliked dogma, Goethe was influenced by dogmatic debates, particularly by those he met, such as Werner in 1787, and Horace-Bénédict de Saussure during his visit to Switzerland in 1779 (Horn et al. 2001, p. 90). One particular influence on Goethe's thinking was German geologist Karl Wilhelm Nose (1753–1835) and his work on basalt.[70]

In reviewing Nose's work, Goethe asked why basalt alone should not follow the temporary development of rock, from oldest to newest?[71] Goethe's morphology calls for some ontogenetic development: a seed becomes a shoot, becomes a leaf, becomes a flower and so on. If basalt can form at any given time, then no such morphological development exists in geology, meaning the same rocks may form randomly at different times within no developmental order. Goethe would agree that we can see an order within the sequences of rocks, in the same way we "see" (read "model") the order in which boulders topple from a tor. These stratigraphic sequences in which granites form at the bottom and rocks form on top can only be reasoned from a Neptunist point of view, namely that these rocks crystallised out of a liquid and settled on a seabed, in the same way sediment does in a lake or shallow sea. "If the basalt was to be volcanic, then foundations had to be sought from which it could arise".[72] Where were these foundations, namely the volcanoes, and flames that created the lava? One way that Neptunists thought rock could melt was via burning coal, such as the burning coal-pits at Dutweil that Goethe had witnessed during his student days.

One of Nose's comments may have struck a chord with Goethe, namely "the historical finger points to the need, to the necessity, of finding out and specifying an original parent rock [Muttergestein] for each volcanic product".[73] For Goethe, Nose's "Muttergestein" may have translated as "Urgestein" or granite. The idea that basalt originates from a parent rock fits in with Neptunism.[74] So does the next passage that Goethe cites.

> Take a good example of another mineralogical process. For as certain plants and animal forms can be found in the rocks and mountains, and if this was to be scientifically considered, the vegetable and animal organisms are equally regarded as foundations, which have been absorbed by the mineral kingdom, assimilated to it, and thereby changed, i.e., petrified [fossilised] in general.[75]

Nose's argument that basalts, like fossils, have a foundational element, such as a parent rock, is born from the existence of altered basalts; that is, basalts that have undergone some form of metamorphosis due to heat. This is not the mysterious internal heat of the Earth or "Hitze" that the Plutonists, such as Hutton, referred to, but rather the heat of fire or fire-effect, like the flames emanating from the burning coal-pits at Dutweil. If burning coal can produce such flames, then a Neptunist interpretation of basalts is valid, rather than more violent igneous sources suggested by the Plutonists. Nose (1820) continues and introduces his idea of pyrotechnics [Pyro-Technik]:

> As is well known, that basalt is more or less meltable: one more so, the other less so, i.e. after mixing the bulk. If it is now subjected to heat, from the red and white glow onwards to melting and liquefaction, one observes the graduated transformations which it thus obtains; this therefore results in a certain standard according to which every fire-effect [Feuer-Effekt] on it is to be precisely determined and correctly judged, and which is especially suitable for adapting it to the phenomena of those basalts which nature allows to be found in certain regions of the world.[76]

Geologists, such as James Hall (1761–1832) and Gregory Watt (1777–1804), had previously melted basalt in an iron furnace to prove Hutton's theory that basalt crystallises from volcanic melt (Watt 1804; see Sosman 1916). So too did Saussure (1779), in his *Voyages dans les Alpes*. Heating rock, melting and allowing rock to recrystallise does not prove an association with volcanoes. Goethe did acknowledge the "close relationship between basalts and volcanoes" but insisted that "basalts were the products of a universal volcanic sea",[77] and that Neptunists have accepted that lavas are molten basalts. One way to show that basalt is formed from a parent rock would be to melt or partially melt an existing rock.

Goethe's experiments into melting rock are perhaps the most underreported. Fortunately they were mentioned by Semper (1914) and more recently and in greater detail by Horn et al. (2001). Goethe was driven to these experiments through his seeming obsession in trying to understand the volcanic nature of the Kammerberg (or Kammerbühl) [modern-day Komorní Hůrka] near Eger [modern-day Cheb] in Bohemia [modern-day Czechia]. Goethe visited the Kammerberg in 1808, 1820 and 1822, each time puzzling over the local geology. The Kammerberg is a stratovolcano, Quaternary in age, with a monogenic scoria cone and interbedded lava flows of basalt and volcanic ash. In his first account of the Kammerberg, Goethe noted the "volcanic products" produced by the hill and noted "[t]he question can arise whether [the Kammerberg] is a real or pseudo-volcano; but whatever view you tend to side, so many things will remain problematic because of particular circumstances".[78]

Goethe's view was undecided. Was this a real volcano that formed in the ocean from an early melt, or was it formed by the combustion of coal, creating the fire-effect that melts parent rock into basalt? It is hard to know what Goethe thought in 1808 as "[i]n the version of 1820, the following two sections have been added".[79] Goethe, like many naturalists of his day, and like many scientists now, could change his mind, particularly over a period of 18 years. How his ideas developed (and conflicted) are discussed in Horn et al. (2001).[80] The final word Goethe had on the matter is evident in the title of his 1824 essay *Wolfsberg, Rehberg and Kammerberg as Pseudo-volcanoes* [*Wolfsberg, Rehberg und Kammerberg als Pseudo-vulkane*][81] (LA I 2, pp. 329–331). Goethe employs his Morphology to interpret the development of these pseudo-volcanoes, particularly the Kammerberg.

> We search for the archetypal rock in those solid basalt rocks; we assume that slate [Tonschiefer] and coal [Steinkohlen] were mixed and pushed into each other; this mixture, subsequently ignited, has melted [verschlackt] and been altered, while still remaining in the same position of its former stratification, but the embers have also seized the adjoining basalt rocks and had a strong influence on the upper layers of the formation, whereas the lowest layers are in their fixed archetypal condition.

The coal, crucial to the pseudo-volcanic hypothesis, was missing from all the volcanic sites in Bohemia. Goethe had a solution:

> If one now further considers that such appearances [volcanoes] in Bohemia, whose pyrotypic characteristics cannot be denied, are to be found on the outgoing [Ausgehenden] coal and brown coal deposits, one would probably be inclined in the end to address all these phenomena as pseudo-volcanic.[82]

Goethe's Neptunism was never in question; however, placing the theory (i.e., Neptunist pseudo-volcanoes) before the evidence (i.e., presence of coal) was questionable. By conveniently explaining the absence of coal on its total consumption by fire seems almost dogmatic, something we will return to later.

Goethe's earlier experiments did reveal some corroborating "evidence", particularly in the recovery of minerals from the basalt, namely that the hornblende crystals survived unaltered from the fire. In this sense Goethe argued, basaltic hornblende is the parent rock [*Urgestein*] and the surrounding slate the archetypical rock. Seen in context to Goethe's Morphology the slate was laid down and remained unaltered in its archetypical condition, while the melting of the younger layers of slate revealed the parent rock and its constituent minerals. How did Goethe arrive at this particular notion, given that his Morphology requires some degree of observation (Anschauung)? "Observation [Anschauen]", Goethe reminds us,

> suddenly gives us the perfect idea of something accomplished; the power of thought, which is also proud of itself, does not want to lag behind, but wants to show and interpret in its own way how it could and had to be done.[83]

The problem is that observation of something that happened in the past, or is hidden underground, is impossible. Here, Goethe reminds us where our imagination comes to the rescue, "... gradually such beings of reason [Gedankenwesen] (*entia rationis*) arise to which the great merit remains of reducing us to contemplation and of urging us to greater attention, to perfect insight".

That perfect insight was the Urphenomenon, but here it is missing, as we are now reduced to using our imagination. What Goethe did do was try to and recreate the fire that he believed melted the parent rock. These were Goethe's "Pyrotechnical experiments", which Horn et al. (2001) divided into "three phases corresponding to his field excursions and sampling 1807/8, 1820 and 1822/23" (Horn et al. 2001, p. 87).

The point of the experiments, carried out in his 1808 *Der Kammerberg bei Eger*, "was made in order to show all the more clearly that the more or less reddened mica schist in the strata of the Kammerberg had passed through a strong fire".[84] The list of 25 specimens includes 19 that were fired in a porcelain kiln [Porzellanofen]:

> 7. Mica schist without quartz, from layers of the Kammerberg. However, its colour is grey and unchanged.
> 8. The same is caused by the porcelain fire, whereby it has become reddish [...]
> 21. Perfect slag.[85]

Horn et al. (2001) summarise Goethe's experiments in greater detail. During Phase I in 1808, mica schist from Kammerberg was fired. Porcelain kilns of that era reached a temperature between 1000°C and 1460°C; however, the specimens Goethe used, housed at the Museum for Mineralogy at the Institute for Geosciences at Jena, showed that temperatures between 1380°C and 1436°C were reached for "a 6 [to] 12 hour heating period" (Horn et al. 2001, p. 89).

Most of what we know of the Phase II experiments in 1820 is listed in *Zur Geologie, besonders der Böhmischen* [*On the geology of Bohemia in particular*]: "6. Pyro-technical experiments by nature and chance, so-called pseudo-volcanic products near Karlsbad [Karlovy Vary] [...] 7. Pyro-technical experiments purposefully by pottery fire",[86] and in various reports, such as "Earth-fired products from the hills over Lessau [Sadov], on the highway to Schalkenwerth [Ostrov] in April/May 1820". In the report, Goethe describes the results of 21 specimens that were fired. Most entries don't identify the parent rock or mineral: "1. unrecognisable rock", while others only give a vague account of what resulted: "yellow porcelain jasper with bluish, uneven and cracked". Other entries are far more revealing: "basalt with augite crystals, burned out". In some entries he identifies the locations where these experiments took place, "the same at the village of Poppen [possibly Březová] between Zwodau [Svatava] and Mariaculm [Chlum Svaté Maří, all locations in the Sokolov District, Karlovy Vary, Czechia]"[87] (LA I 2, pp. 141–142). The detailed studies of Semper (1914), Horn et al. (2001) and Prescher (1978) reveal little about how Goethe conducted his experiments in Bohemia, simply because so little of what he did was recorded. All we have are lists of rocks that had been fired, such as "pseudovolcanic products from Hohdorf [Dalovice] with red numbers" (LA I 2 pp. 142–143), "following quartzite into conglomerate rock. With large red Latin letters" (LA I 2 p. 143) and "passed through the porcelain fire Bergarten. With small red Greek letters" (LA I 2. pp. 143–144).[88]

The letters and diary entries also fail to reveal any method other than the firing of various rocks. Horn et al. (2001) did find a notable summary of Goethe's experiments in 1820.

It is a scientist's first duty, when studying volcanoes and Earth fires, to investigate if he can discover the original rock type from which the altered ones originated. We have studied this more carefully than normally and therefore a large number of rock types have been tested in the kiln fire, and remarkable observations of "obstinate" and of easily affected rock have been made.

We have made an ordered collection of these where the originals and altered have been juxtaposed and marked by numbers. It would be useless and verbose to publish and describe these objects fully as they are; we shall therefore, after the hints we got by experience, go on, and with the help from Court Council Döbereiner, endeavour to present the instructive aspects in systematic order.

(LA I 2 p. 163, translated in Horn et al. 2001, pp. 92–93)

Goethe's last pyrotechnic experiments were conducted in 1823. The rocks in this experiment were sampled from Wolfsberg [Vlçi Hora, Šluknov Hook, Czechia]. Goethe's aim, again, was to find the parent rock that had been altered. In these experiments, Horn et al. (2001, p. 93) note, Goethe found that the pyroxene mineral augite "remained unaltered", indicating that it may have been part of the parent rock. In other words, if we melt basalt we should recover the minerals of the parent rock gneiss, which Neptunists believed formed after granite. High metamorphic grades of gneiss contain augite; however, the pyroxene is more commonly found as phenocrysts in basalt and ultra-mafic intrusive rock such as gabbro.

Much of Goethe's Morphology is seemingly absent from his work on basalt. His misreading of Nose's pseudo-volcanic rocks as Neptunistic (*sensu* Semper 1914), and his experiments attempting to find the original parent rock, all have the hallmarks of trumping evidence. Goethe's experience of the burning coal-pits in Dutweil, the positioning of lava with the altered parent rocks on the Kammerberg and his genuine belief in Neptunism all conflate with his Morphology, namely Anschauung and the Urphenomenon. Yet, in his appraisal of Nose (1820), Goethe muses,

… the most important question appears: to what extent we may declare an unexplored thing to be inscrutable, and how far is it permitted for man to go forward before he has reason to recede before the incomprehensible or to stand still before it? Our opinion is: that man is expected to assume something unfathomable, but that he has no limit to set his inquiry; for even if nature has an advantage over man, and seems to conceal many things from him, he again has an advantage over her, so that he can think beyond her, even if he passes through it. We have already advanced far enough against them when we reach the Urphenomena, which we look at face to face in their inscrutable glory and then turn backwards again into the world of phenomena, where what is incomprehensible in its simplicity reveals itself unchangeably in thousands and thousands of manifold phenomena despite all their changeability.[89]

But where and how was this applied in his pyrotechnical experiments? Seeing rock melt into slag isn't comparable to the formation of basalt by underground coal formations burning parent rock. In 1820, and certainly before, Goethe succumbed to Neptunistic dogmatism. Almost a decade later he renounced Werner's teaching as dogmatic, and that "all the dogmatism in the world becomes annoying in the end" and "leaves behind unexplained problems".[90]

Goethe pondered basalt and its formation for a great deal of time, leaving many unexplained problems. It is clear, in hindsight, that Goethe's Morphology didn't work in geology, as his experiments in basalt have shown.

KARLSBADER ZWILLINGE

In 1807 Goethe composed a poem titled "The Mountains of, and around, Carlsbad [Karlovy Vary, Czechia]. A draft" [*Gebirge von und um Karlsbad Entwurf*], in which he described a porphyritic granite:

Rhomboid pieces of feldspar.
Sharpened and merging into crystal form.
Double or twinned-crystals.
Visible in many steps on the houses.
After granite has dissolved, in certain places, they appear as strange remnants.
Of the same form.
Double crystals.
Washed on top and on each other.
Created in the cross.[91]

Goethe has used his Morphology to optimal effect in describing the twinning of the potassium feldspar mineral orthoclase.

Goethe had described Carlsbad twinning[92] in a way that is easily observable from any angle, without deferring to Werner's Neptunism (Figure 2.4).

In his description of the rock and mineral collection of Joseph Müller,[93] Goethe described the granite and the feldspar crystals using Morphology.

> Large rhombic feldspar pieces are noticeable in the coarse-grained specimens [of granite]. Their internal structure and external form indicate a crystallisation, which expresses itself more decisively. There are large masses of Carlsbad granite in which one finds complete crystals, and indeed of a very complicated formation [Bildung] (5). They are double crystals, which seem to consist of two crystals interlocking and overlapping each other, without one being able to imagine one without the other. Their form cannot be brought forward to the imagination by description, but one can imagine them as two interlocking rhombic plates (6, 7, 8).[94]

The last line is rather telling. "Their form cannot be brought forward to the imagination by description [Form ist durch Beschreibung nicht wohl vor die Einbildungskraft zu bringen]" is perhaps the crux (no pun intended) of Goethe's *Morphology*. You need to imagine the form via Anschuuang and the Urphenomenon and spin the Carlsbad twin around in your mind. There is an actual eyewitness account of Goethe doing this in an extract titled *Mineralogists Memoir*[95] by German mineralogist Christian Samuel Weiß (1780–1856). In the summer of 1818 ("August/September"), Weiß met Goethe in Carlsbad to discuss feldspar crystals, in particular the Carlsbad twin. Weiß noted, "that, for the sake of brevity, [Goethe] preferred to call them Carlsbad twins". After attempting to describe the twinning to Goethe, Weiß wrote:

> With a very satisfied and comfortable 'Hm! Hm!' he rocked his head forwards and backwards; then he had his hands open a little, and in describing this manoeuvre, he said: 'So some of them are like that! In the others like that!' And with that he clapped his hands together, at first in the conventional manner, with the inner side (the palms of his hands) against each other, then, after he had turned them over, both again with the backs of his hands against each other. That was the entire manoeuvre needed to make his point. And I, not a little taken aback at first, only realised after thinking about it for a while how apt his figurative representation was.[96]

To a modern eye this is quite a good way of describing Carlsbad twinning figuratively. Joining both hands together (palm to palm, or in prayer) shows perfect symmetry, as each palm, fingers and thumb matches the other. Figuratively each hand is a tabular piece of orthoclase, which, joined, forms what is known in mineralogy as simple

FIGURE 2.4 Karlsbader Zwillinge. (Drawing possibly by Goethe during 1806–1807 (LA I 1, plate XVI).)

twinning or simple twins. By turning the hands around so the backs of the hands are touching is quite cumbersome. The action means the arms are twisted and no perfect symmetry between hands is achieved. Try it and you will see that both hands will be at angles to each other, an excellent figurative example of Carlsbad twinning.

ON GOETHITE

In a letter to Ludwig Wilhelm Cramer (1755–1832) Goethe asked to pass his thanks on to Dr Heinrich-Adolf Achenbach (1765–1819).

> Please send my greetings to Pastor Achenbach [...] whether it has just lost its name, Goethite will always be dear to me.[97]

Writing in *Zur Morphologie* in 1820, Goethe recalled at being humbled by having a mineral named after him.

> Benevolent men in the Westerwald discovered a beautiful mineral and named it Goethite in my honour; I am indebted to Messers Cramer and Achenbach.[98]

The naming and usage of Goethite [Goethit] is part of a convoluted story. The mineral was found in 1806 near Siegen by Johann Daniel Engels (1761–1828), the Siegen Bergmeister [mine manager] who in consultation with his friend and mentor Reverend Heinrich-Adolf Achenbach (1765–1819) called upon mineralogist Christoph Ludwig Wilhelm Cramer (1755–1832) to name it "Goethenit" (Schroeder 1988, p. 138). The specimen made its way to Johann Georg Lenz (1748–1832), Professor of mineralogy at the University of Jena, who, after consulting with Friedrich Wilhelm Riemer (1774–1845), named it Göthit [Goethite]. Lenz (1806) described it in his *Tabellen fur das gesamte Mineralreich* as,

> this dark ruby-red, three-sided tabular crystalline form of hematite [Specularite], from the Eisenzeche [mine] near Siegen is quite brittle [...] and is named in honour of our most dignified president, the Privy Councillor von Goethe.[99]

The mineral, however, had already been discovered and named Pyrrhosiderit or Rubinglimmer (a compound term consisting of Rubinrot and Eisenglimmer) by Johann Christoph Ullmann (1771–1821), a professor of mineralogy at the University of Marburg. Ullmann had been using the term Pyrrhosiderit in class for years, something of which Achenbach and Engels were clearly unaware. Ullmann eventually published the term Pyrrhosiderit in his Systematic overview of minerals [*Systematisch-tabellarische Uebersicht der mineralogisch-einfachen Fossilien*] in 1814, as "ruby-red Specularite; *Becher*. (Lepidocrocite). Goethite; *Lenz*".[100]

Over time Lepidocrocite [Rubinglimmer] became the preferred term, appearing in the 1826 edition of the *Handbook of Mineralogy* [*Handbuch der Oryktognosie*] by Karl Cäsar von Leonhard (1826) as:

> 1. Brown Hematite.
>   a. Lepidocrocite.Pyrrhosiderite, Goethite, fibrous Brown Hematite, *Fer pourpré* ....[101]

Goethe lamented that the name Goethite wasn't officially adopted,

> … I am indebted to Messers Cramer and Achenbach [for naming Goethite], even though the name has disappeared from mineralogy [Oryktognosie]. [Goethite] is also called Lepidocrocite, and at the moment it is known as Pyrrhosiderite. For me it was enough that you have thought of naming such a beautiful natural product after me, even if for a moment.[102]

There is, however, a twist to the story. In 1901 François Alfred Lacroix (1863–1948) examined the original specimens from Siegen and noted that Goethite and Pyrrhosiderite were both gamma forms of iron oxide-hydroxide [γ-$Fe^{3+}O(OH)$],[103] now known as Lepidocrocite. Lacroix (1901) concluded that Lepidocrocite had priority as the name for the gamma form (given Pyrrhosiderite was introduced 8 years later by Ullmann in 1814), but instead assigned the name Goethite to the alpha form [α-$Fe^{3+}O(OH)$], for which it is known today.

ON THE ICE AGE

Goethe's fragments on the *Eiszeit* or Ice Age[104] start off with a noteworthy,

> Ice plays a greater role than we think.[105]

The fragments Goethe wrote on the "epoch of great cold" are scattered throughout his geological writings: "*Herrn von Hoffs geologisches Werk*", written in 1823, "*Geologische Probleme und Versuch ihrer Auflösung* [1829], *Kälte* [1829], *Gebirgs-Gestaltung im Ganzen und Einzelnen* [1824], *Umherliegende Granit*, *Erratische Blocke* [1829 …] and *Eiszeit* [1829]" (Cameron 1965, p. 751), as well as *Wilhelm Meister's Journeyman Years* (1821). Perhaps, Wilhelm Meister provides a better narrative for Goethe's ideas on glaciers. Wilhelm, invited to a miners' festival, reconnected with Jarno (aka Montan). After the ceremony, the guests seated themselves to dine, engaging in conversation about "mountains, views and deposits, of the types of views in the region" (Goethe 1989, pp. 278–279). A Neptunist–Vulcanist debate ensued about the formation of mountains and volcanoes. The discussion shifted to the explanation of erratic boulders, which one group explained to have been distributed via "convulsions" brought upon by "elastic forces" within the earth's crust.

> A fourth, if not perhaps numerous, party, scoffed at all these futile attempts and maintained that many features of the earth's surface could never be explained unless it was allowed that greater and lesser segments of mountain could have fallen from the sky and covered broad stretches of the landscape […]. Finally two or three guests invoked a period of fierce cold, when glaciers descended from the highest mountain ranges far into the land [northern German plains], forming in effect slides for ponderous masses of primeval rock [granite], which were propelled farther and farther over the glassy track. In the subsequent period of the thaw, these rocks had sunk deep into the ground, to remain forever locked in alien territory. In addition, the transport of huge blocks of stone from the north [Scandinavia] might have been made possible by moving ice floes.
>
> *(Goethe 1989, p. 279)*

Goethe wrote this in 1829,[106] almost a decade before Louis Agassiz (1807–1873), Karl Friedrich Schrimper (1803–1867) and Jean de Charpentier (1786–1855) proposed the idea of an "ice age".

Agassiz, considered to be the founder of the ice age, attributed the idea to Goethe who "alone unified all the indications into a definitive theory" (Cameron 1965, p. 751). Goethe's ideas on the origins of erratic boulders did significantly influence his art, in the case of *Wilhelm Meister's Journeyman Years*, as well as his science. The narrative in *Wilhelm Meister's Journeyman Years* conveys an overall idea, but it lacks detail, something we do find in Goethe's fragments, such as the following examples.

> Big ice rafts carry granite blocks in the Baltic.[107]
>
> For so much ice to exist, cold is needed. I suspect that there was an epoch of great cold in Europe.[108]
>
> The glaciers travel through the valleys to the edge of the lake carrying the granite blocks loosened from above, as still happens today. The blocks remain on the lake plain after the ice melts, to be found today, unrounded, because they were brought there smoothly, and not forcefully.[109]

Unlike his narrative in *Wilhelm Meister's Journeyman Years* these fragments introduce geological principles, such as roundness, which indicates transportation distance. Another idea independently derived by Goethe was the "epoch of great cold", a cold climatic period in which the northern German plains were glaciated, as opposed to the extension of existing glaciers, as championed by Charpentier (see Cameron 1965).

ON METEORITES

Goethe had several meteorites in his collection: the Ensisheimer meteorite from Ensisheim, Haut-Rhin department, France, a gift from Johann Friedrich von Blumenbach (1752–1840) sent on 20 August 1803 (Prescher 1978, No. 1155, p. 568, footnote 40); a meteorite from Limerick in Ireland (Prescher 1978, No. 6605); three pieces of meteorite which fell in Pohlitz near Gera, Thuringia State, Germany, on 27 August 1819 (Prescher 1978, No. 6968); and a meteorite which fell in Ruden in Kurland, modern-day Latvia on 31 January 1686 (Prescher 1978, No. 7801).

Goethe thanked Blumenbach for "the sample of atmospheric rock",[110] which he referred to again 18 years later in a letter to Austrian naturalist Carl Franz Anton von Schreibers (1775–1852).

> It is a very peculiar feeling when I now place both of these creations [Werke] I have just reported next to each other and see in both incomprehensible natural products before me. Plants, of which it is impossible to foresee how they could grow near the fiftieth degree latitude; Minerals that suddenly rush down out of the blue, where one would not expect such a thing, in a concentrated manner. We are happy to let all this happen as soon as the attentive observer wants to present us with such phenomena in a great general context and make them clear, whereby alone the mind [Geist] can be set free against such tremendous objects.[111]

Freed from the bonds of Neptunist thought, Goethe suddenly realises how little he actually knows. Two familiar natural objects become incomprehensible when they are found to be in areas where you don't expect them to be; be it an earthbound rock falling from the sky or a tropical plant living in cool temperate climes. The Ensisheimer meteorite is an ordinary stony chondritic meteorite, but to an early 19th-century observer it looked like breccia, a familiar sedimentary rock. Where would a meteorite fit into Werner's classification? Goethe never reveals his thoughts on an actual observed geological event. Regardless, Goethe continued his Wernerian slumber, unaware, perhaps, that this was as close as he was going to get to the real Urgestein.

ON GRANULITE

The term granulite first appeared in the geological literature in Weiss (1803).

> Nor can I deny that I myself do not like to speak of a grainy white-stone, nor of newer granite and white-stone. Therefore, if I were allowed to take a middle path and suggest a name that would fit both variations, it would probably be granulite.[112]
>
> *(Weiß 1803, p. 348)*

Granulite is a high-grade metamorphic rock[113]

> in which Fe–Mg-silicates are dominantly hydroxy-free; the presence of feldspar and the absence of primary muscovite are critical, and cordierite may also be present. The rock with >30% mafic minerals may be called mafic granulites, and those with <30% mafic minerals (dominantly pyroxene) may be called felsic granulites.
>
> *(Coutinho et al. 2007, p. 56)*

In 1784 Goethe noted an odd rock he collected during his third Harz trip.

> 22. Strange rock from the Bude Valley, diagonally opposite the Susenburg [Suseburg], which I dare not classify as either granite or porphyry, and for which I suggest the name granulite, because of the quartz grains found in them. It is extremely solid.[114]

Kuhn and von Engelhardt note that,

> The name proposed by Goethe for a *strange rock* does not correspond to the modern name, which goes back to Weiß [1803], for a light-coloured, schistose rock made of quartz and feldspar with garnet as well as kyanite [Disthen] and/or pyroxene.[115]

The rock identified as granulite by Goethe in 1784[116] could be a kyanite schist, a medium- to high-grade metamorphic rock and is definitely not a granulite (*sensu* Coutinho et al. 2007). The rock proposed by Weiß (1803) is still considered to be a granulite and is part of the Saxonian Granulite complex in the Granulite mountains [Granulitgebirge], near Waldheim, Germany (see Borghini et al. 2019).

GEOLOGICAL MAP OF GERMANY 1821

In 1821 Christian Keferstein (1784–1866) published the first geological treatment of Germany,[117] which he dedicated to Goethe "with adoration and gratitude".[118] The volume came with the first geological map of Germany [*General Karte von Deutschland*] (Figure 2.5). In the preface Keferstein notes:

> In the illumination of the charts, it was not only a matter of indicating different formations by different colours, but also of choosing the colours in such a way that they are conveniently distributed and pleasantly conspicuous; in this the assistance of a great connoisseur of colours was of great importance. Privy Councillor von Goethe was so kind as not only to take an interest in this work in general, but also to indicate a colour chart which will form the basis of the illumination of the entire Geognostic Atlas, and I take this opportunity to repeat my heartfelt thanks for it here.[119]

FIGURE 2.5 Geological map of Germany. Germany presented geologically, with maps and cross-sectional drawings, which form a geological atlas. [*General Karte von Deutschland.* Deutschland geognostisch-geologisch dargestellt, mit Karten und Durchschnittszeichnungen, welche einen geognostischen Atlas bilden, Fig. 1.] Published in 1821 by Christian Keferstein.

Geological maps, including those of today, are coloured, with each colour and hue indicating a rock type. Goethe had proposed a consistent colour scheme, something that,

> [i]f the intended geognostic atlas is carried out in this way, it would be desirable that the connoisseurs of this science would unite and use the same colours to designate the same rock, from which a quicker overview would emerge and many a convenience would arise.[120]

The colour scheme proposed by Goethe was based on "the old petrographical classification of rocks according to their nature"; however, none were "decisive in Goethe's colour selection" (Schäfer-Weiss and Versemann 2005, p. 170). In a review of Keferstein's volume, Goethe outlines his reasoning for the selection of colour found in the legend titled *Explanation of Colours* [*Farben Erklärung*].

> The main formation, which contains granite, gneiss, mica schist with all its deviations and deposits, is given the carmine colour, the purest most beautiful red; the immediately adjacent slate was given the harmonising pure green; on top of the alpine limestone the violet, also related to the red, not reluctant to the green.[121]

Schäfer-Weiss and Versemann (2005) noted the similarities between Goethe's review and his opening paragraph of *Effect of Colour with Reference to Aesthetic and Moral Associations*,[122] highlighting the importance of his *Farbenlehre* in the colour selection found in the *Karte*. Given this, would it be prudent to ask whether Goethe's "Explanation of Colours" was one of the lasting legacies of his work on geology? Colourised geological maps were published before the *Karte*.[123] While Semper (1914) and von Engelhardt (2003) agree that it was, and Steiner (1957, 1969) and Wagenbreth (1967) believe it was not, claiming Werner's colour system as the one still used today. Schäfer-Weiss and Versemann (2005), however, take a more nuanced view, namely,

> Neither Werner's colour chart nor Goethe's provided the entire basis for the colouring of modern geological maps. Instead, individual components from each of their schemes have been adopted .
>
> *Schäfer-Weiss and Versemann (2005, p. 173)*

GOETHE'S GEOLOGICAL LEGACY

Goethe's geological legacy, both in his observations of rocks, particularly granite, and his interest in cartography, had a major impact on his later works, such as *Wilhelm Meister's Journeyman Years*, *Elective Affinities* and *Faust II*.[124] The impact Goethe's work had on practical geology is very limited. He may have formalised geological terms, in use at the time in mining, his work on granitic tors and helping to standardise the use of colouring in geological maps. Much of his work was mostly a narrative account of his travels and encounters with outcrops or lists of specimens he had collected. He had speculated on the origins of minerals and rocks, their stratigraphy and the origin of the earth. Most of these speculations were constrained by his unshakable belief in Werner's Neptunist system.

Goethe's Morphology, his universal method, failed in his geological investigations; however, his fear that "… without [geology] the famous Morphology would never be complete" is overstated. Morphology does not lend itself to unobserved processes. Goethe classified rocks and minerals by unobservable Neptunist processes, grouping rocks we now know to be petrologically and mineralogically unrelated, such as granite and gneiss.[125] Without a larger suite of specimens and access to better microscopes, Goethe had little chance to observe the minerals and foliations to link, say, a slate to a phyllite and a phyllite to a gneiss. Even if he did, Goethe was unable to observe one becoming another, in the same way he could observe the metamorphosis of a leaf. Goethe's main problem was his adherence to dogma or synthesis that was unobservable.[126] Neptunism, or any speculative ideas for that matter, does not allow for Urphenomenon to arise through Anschauung, simply because ideas are unobservable. Morphology relies on something to be observed, namely form, and as Elizabeth Mary Wilkinson once noted, "Form is that which has actuality" (Wilkinson 1951, p. 186).

Perhaps we can be more blunt and say, "Without form, there is nothing".

NOTES

1 The source of this quotation, originally written in English, is unknown (see Hennig 1949).

2 "Wir hörten von den reichen Dudweiler Steinkohlengruben, von Eisen- und Alaunwerken, ja sogar von einem brennenden Berge, und rüsteten uns, diese Wunder in der Nähe zu beschauen …" (LA I 1, p. 2).

3 "In der Alaunhütte erkundigten wir uns genau nach der Gewinnung und Reinigung dieses so nötigen Materials, und als wir große Haufen eines weißen, fetten, lockeren, erdigen Wesen bemerkten und dessen Nutzen erforschten …" (LA I 1, p. 3).

4 … Wir traten in eine Klamme und fanden uns in der Region des brennenden Berges. Ein starker Schwefelgeruch umzog uns; die eine Seite der Hohle war nahezu glühend, mit rötlichem weißgebranntem Stein bedeckt; ein dicker Dampf steig aus den Klüsen, und man fühlte die Hitze des Bodens auch durch die starken Sohlen. (LA I 1, p. 3)

5 "Als Werner'scher Schüler konnte sich Goethe mit den Lehren der Plutonisten nicht befreunden, wenn er auch die unhaltbare Stellung der schroffen Neptunisten wohl durchschaute" [and] "Von nennenswerther Bedeutung sind übrigens Goethe's Leistungen auf dem Gebiete der Geognosie nicht" (von Zittel 1899, p. 99).

6 *Verhältnis zur Wissenschaft, besonders zur Geologie* (LA I 2, pp. 147–150; see also Goethe 1995, pp. 138–140).

7 Most notably his attempt to console the Plutonists and Neptunists on the origin of basalt (1789), his anti-vulcanism, his interaction with German-speaking geologists, as well as the development of his impressive mineral and rock collection in Weimar (Prescher 1978; Hamm 2001).

8 Prescher (1978) is the antidote to the first two works of the *Leopoldina Ausgabe* (LA), namely LA I 1 and LA I 2. Both these works simply list Goethe's geological works in chronological order without an index. Each volume only contains a table of contents. In contrast, the catalogue of Prescher (1978) is invaluable, as it indexes all author names, geographical locations and object names, as well as providing a comprehensive list of Goethe's entire collection of rocks, minerals and fossils, which are cross-referenced to previous catalogues. These numbers are linked to the specimens discussed in Goethe's works. Prescher (1978) is a godsend to anyone attempting to dredge through the first two volumes of the LA.

9 Wells notes that Goethe's interest in geology was reawakened after Schiller's death (see Wells 1978, p. 55).

10 Yet, Caisley (2020) notes,

> There are some strikingly detailed studies, such as Engelhardt's exceptionally thorough survey of Goethe's written output in the sphere of geology and Semper's chronological review of the development of Goethe's thought in this field, but neither Engelhardt nor Semper undertake any deep analysis of the texts rather than simply repeating Goethe's statements within them. Indeed, there is next to no scholarship that explores Goethe's written output with regard to the impact that Ilmenau had on his thinking.
>
> *(Caisley 2020, p. 138)*

11 See Chapter 1.

12 Geognosy is synonymous with geology. It is defined as "[T]he description and classification of the rocks and rock formations (and, in later use, other inorganic substances) of which the earth consists; the branch of natural science (or later of geology) dealing with this". From: "geognosy, n.". OED Online. June 2020. Oxford University Press. https://www.oed.com/view/Entry/77749?redirectedFrom=geognosy (accessed August 14, 2020).

13 "Es ist mir sehr lieb, daß ich so zufälligerweise diese Betrachtungen erneuert habe, ohne welche denn doch die berühmte Morphologie nicht vollständig werden würde" Goethe to Schiller 12.11.1796 (WA IV 11, p. 260).

14 Due to Goethe's association with Neptunism, he "has not had a lasting influence on geology" (Baldridge 1984, p. 163). Wells (1978) also believed that Goethe's "genetic relationships" [part of his Morphology], which worked well in biology, didn't work in his "geology investigations" (Wells 1978, p. 72).

15

> Der Granit is durch Kristallisation entstanden.
> An ihm ist keine Gravitation zu bemerken.
> So auch die nächsten Gebirge an ihm.
> Je weiter es von ihm weg kommt, je mehr nimmt die Schwere überhand, bis zuletzt bei den Flötzen nur eine Spur von Kristallisation übrig bleibt.
> Der ganze Bau unserer Erde ist aus der Kristallisation zu erklären.
>
> *(LA I 1, p. 98)*

16 Kurze Klassifikation und Beschreibung der verschiedenen Gebirgsarten (Werner 1787).

17 "Diese sind: urfängliche-, Flöz-, vulkanische-, and aufgeschwemmten Gebirgesarten" (Werner 1787, p. 5).

18 Die uranfänglichen-, Flöz-, and aufgeschwemmten Gebirgsarten, gehen in einander über, und man hat uranfänglichen Gebirgsarten, die beinahe mit eben dem Rechte auch Flözgebirgsarten genennt werden können; und wiederum Flözgebirgsarten, wo man zweifelhaft wird, ob man sie nicht schon zu den aufgeschwemmten zählen soll. Einige der ersten, nämlich der uranfänglichen, gehen auch ganz allmählig in Flözgebirgsarten über. Nach den Entstehungsarten dieser Gebirgsarten, die sich in dem ungeheuren Zeitraume der Existenz unsere Erde wohl meist unmerklich eine in die andere umänderten, ist es auch nicht anders möglich, als daß solche Übergänge bei diesen Gesteinarten statt finden müssen. (Werner 1787, p. 5, my emphasis)

19 "I first composed this Treatise on Rocks based on several German works" ["J'avais d'abord composé ce Traité des Roches d'après plusieurs ouvrages allemands …"] (Brochant 1803, p. 559).

20 Graywacke [*Grauwacke*] was named Rock of Year for 2023 [Gestein des Jahres 2023] by the Berufsverbands Deutscher Geowissenschaftler (BDG): https://www.bv-miro.org/aktuell/grauwacke-goethe-und-das-gestein-des-jahres-2023/

21 "Dieses sogenannte Übergangsgebirge ist mir höchst merkwürdig, weil ich seine Eigenthümlichkeiten erst jetzt kennen lerne, der ich sonst nur im Ur- und (in seinem Gegensatz) dem Flötzgebirge verweilte" (To August Goethe, 8 June 1815, WA IV 26, p. 8).

22 The Biogenetic Law (Haeckel 1866) states that ontogeny recapitulates phylogeny, an idea that was trounced by Nelson (1978). Evo-devo refers to Evolutionary Developmental biology.

23 Urgestein has two definitions. The older, and more relevant, definition refers to rocks still in their original states, that is, not metamorphosed (foliated) or degraded in any way. Granite is usually used as an example of a Urgestein. The term is synonymous with "primitive rock" (Urgebirge) and basement rock (Grundgebirge). The modern usage of Urgestein refers to "veterans", as in a "political veteran".

24 "Goethe sah in Granit das 'Urgestein'" (Riedel et al. 2012, p. 8).

25 "So versuchte er [Goethe], durch eigene Thesen zum Granit, in den Streit einzugreifen und behauptete, dass der Granit ein Urgestein darstelle, das allen weiteren Entwicklungen, auch den vulkanischen, zugrunde liege" (Schneider 2021, p. 21).

26 "… daß durch teilweise Auflösung wie teilweise Beharrlichkeit des Urgesteins …" (1820; LA I 2, p. 172).

27 "Wir haben von dem sämmtlichen Urgestein Musterstücke mitgenommen und einen Kasten in Eger stehen lassen" (Goethe to August Goethe, 29 April 1820, WA IV, p. 4).

28 "Ferner hat schon im vorigen Jahre Herr Graf Sternberg die in Marienbad geognostisch beschäftigten Naturfreunde aufmerksam gemacht auf die so langsame als große Gewalt, welche die Gasarten auf's Urgestein ausübe" (Goethe to Kaspar von. Sternberg 10 September 1823, WA IV 37, p. 220).

29 "84. Anstoßendes Urgestein […] 5. Bis zur Unkenntlichkeit verändertes Urgestein mit frischem Bruch […] 6. Desglichen in runder Kugel" (Prescher 1978, Nos. 7320, 7477 and 7478).

30 Or, terms that are attributed to Werner (Ospovat 1959).

31 There are other terms Goethe uses with the prefix Ur, namely, Urbild, Urfrage, and Urpolarität.

32 See Wells (1978, p. 48).

33 "… als Unterlage aller geologischen Bildung" (LA I 1, p. 61).

34

> Granit ein mannigfaltiges Gestein.
> So häufig und doch kaum aufweisen.
> Genau betrachtet. Tritt die Hornblende ein: Syenit.
> Oder er wird schiefrig [sic], geht in Gneis über.
> Oder enthält seine Teile größer, für sich vergrößert,
> porphyrartiger Granit gennant.
> So daß man recht das augenblickliche Werden, das Schwanken, das Übergehen zu beachten genötigt. Echantillons von eigentlichem Granit.
>
> *(LA I 1 p. 347)*

35 Goethe may be referring to Carlsbad twinning, as the rocks he is describing hail from Karlsbad (modern-day Karlovy Vary, Czechia). If so, the feldspar he is referring to is the potassium feldspar orthoclase. According to Christian Samuel Weiß, in 1818 Goethe coined Karlsbader Zwillinge (Carlsbad twinning), typically found in orthoclase (K-feldspar), "… daß er sie der Kürze halber am liebsten Karlsbader Zwillinge zu nennen pflegte" (LA I 2, p. 118, and see below).

36 Jener Granit findet sich auch bei seinem Übergang in Gneis. Jene Zwillingskristalle in ihrer noch völlig entschiedenen Form and Eigenschaft erleiden Einfluß durch den Glimmer, dessen flache Richtung sich hier zu manifestieren anfängt. Sie erscheinen selbst verflechtet, gestreckt, indem der Glimmer in sie übergegangen und auf sie eingewirkt hat. Sie dagegen bestimmen die Gestalt des ganzen Steins, indem das Flasrige desselben ganz allein durch sie hervorgebracht wird. (LA I 1 p. 383)

37 The crystallisation into the Bowen's reaction series places biotite mica before potassium feldspar (orthoclase), but after sodium-rich feldspar, such as albite. So either the rock Goethe is describing has biotite with albite (possibly an intermediate intrusive igneous rock, such as granodiorite), or muscovite mica with orthoclase (possibly a more felsic intrusive igneous rock such as granite).

38 Also published as *Über den Granit* in 1784 (LA I 1 pp. 57–63).

39 Of 17 points under the heading "Rechenschaft wäre zu geben" or "Accountability would have to be given", number 10. "Allgemeiner Typus" or "General Type" might refer to Goethe's attempt to implement the biological notion of the archetype to minerals and rocks (LA I 2, p. 425). The entries are listed in chronological order suggesting the development of Goethe's ideas. Given this, the "General type" is before entry number 11. "Riese auf den Fichtelberg" or Journey to the Fichtel mountains, which occurred in 1785.

40 Goethe vollzieht die temporalisierung der Naturgeschichte mit. Er tut es allerdings nicht primär zur Erweiterung des empirischen Wissens von der Erdentstehung. Statt sich in steingraue Zeiten zu verlieren, versucht er, diese in sinnlichen Maßen unvorstellbaren, darum in Kantischen Sinn erhabenen Kognitionen zu verbinden mit der allerdings erfahrbaren Spannung zwischen unvordenklichem Seinsgrund und pathischer Existenz. Dafür setzt er den Granit nicht als wissenschaftliches Objekt, sondern als symbolische Form ein, nicht *thesei*, sondern *physei*, und bringt ihn derart zur ästhetischen Evidenz. In dieser dialektischen Spannung spiegelt sich bei Goethe die Konfiguration von Geschichte und Natur, von Zeitlichkeit und Seinsgrund, von Subjektivem und Objektivem. (Böhme 2016, p. 57)

41 This idea has been explored by linguists Ferdinand de Saussure and Noam Chomsky.

42 According to Hamm (1991), some historians of geology (Wells 1978; Oldroyd 1979) concluded that "Goethe's geology lacked the concept of time" (Hamm 1991, p. 330).

43 Es unterscheidet sich diese merkwürdige Gesteinart dadurch von allen andern daß sie zwar nicht einfach ist sondern aus sichtbaren Teilen besteht, jedoch zeight der erste Anblick daß diese Teile durch kein drittes Mittel verbunden sind, sondern nur an- und nebeneinander bestehen und sich selbst untereinander festhalten. Wir nennen diese voneinander wohl zu unterscheiden Teile: Quarz, Feldspat, Glimmer wozu noch manchmal einige als Schörl hinzukommen.

Wenn wir diese Teile genau betrachten so kömmt uns vor als ob sie nicht wie man es sonst von Teilen denken muß vor dem Ganzen gewesen sein, sie scheinen nicht zusammengesetzt oder aneinander gebracht sondern zugleich mit ihrem Ganzen das ausmachen entstanden. Und obgleich nur der Glimmer öfters in seiner sechsseitigen, tafelartigen Kristallisation erscheint und Quartz, und Felsdpat, weil es ihnen an Raum gebrach, die ihnen eigne Gestalten nicht annehmen konnten so sieht man doch offenbar, daß der Granit durch eine lebendige, bei ihrem Ursprung innerlich sehr zusammengedrängte Kristallisation entstanden ist. Es sei uns erlaubt auf die Entstehung desselben und auf die Materie woraus er entstanden einige Schlüsse zu machen. (LA I 11, pp. 9–10)

44 Es würde nunmehr kaum Verwegenheit sein, denjeniger Glimmerschiefer porphyrartig zu nennen, welcher durch Quarztelie eine flasrige Textur erhält, denn dieser Quarz ist, obgleich ohne bestimmte Form, doch aber platten- und lagenweis aus der Glimmermasse hervorgetreten, wie man das Umgekehrte ebensogut sagen kann, da in stärken Quarzpartien der Glimmer enthalten ist. (LA I 1, pp. 383–384)

45 *Zur Theorie de Gesteinslargerung* 1785 (LA I 1, pp. 94-98).

46 Early in his *Theory of Rock Formation 1785* [*Zur Theorie der Gesteinslargerung 1785*] (LA I 1, pp. 94-98), Goethe explains the sequence of events regarding the formation of gneiss:

> During the first Epoch a single layer of granite was found all over the world.
> At first a mass of clay [Ton] and mica [Glimmer] were deposited after [the formation of] the granite, which covered it in uneven layers.
> This deposition happened in water. It happened at the shortly after the granite had crystallised, then we find these increasingly growing into the granite, in an alternating way.
> Gneiss is the granite that had fully formed from the water, from which we get its foliated form. The pure argillaceous rocks [e.g., mudstone, siltstone, shale] and therefore the foliated slate or phyllite. Siliceous jasper, when quartz and feldspar have more or less been crystallised.
> [Die erste Epoche des Granits ist einfach und allgemein über die ganze Welt.
> Was sich nach dem Granit zuerst niederschlug, war eine ungeheure Masse von Ton und Glimmer, die überall den Granit in einer gewissen Höhe bedeckt. Diese war gleichfalls sehr allgemein, doch nicht einfach.
> Dieser Niederschlag ist in dem Wasser geschehen. Es geschah gleich, nachdem der Granit sich kristallisiert hatte, denn wir finden diese Gesteigert in den Granit verwachsen, ja mit ihm abwechselnd.
> Gneis ist der Granit, der sich nach der ersten Grundbildung aus dem Wasser Niederschlag, daher seine blättrige Gestalt. Das Tongestein sehr rein und daher blättriger Tonschiefer. Sehr kieselhafter [?] Jaspis, wenn sich der Quarz und Feld[? spat] mehr oder weniger darin kristallisiert].
>
> *(LA I 1, p. 97)*

47 "Bei dem ersten Überschreiten des Granites, also dem ersten erscheinenden Vielfältigen, tritt Kristallisation, also das erste gelungene Individualisieren der Natur ein. Bei diesem Letzten zeigen sich organische Gebilde sowohl großer Farrenkräuter als der Korallen" (LA I 1, p. 381).

48

> Alles Geologische liegt zwischen einem Ältesten und Jüngsten; zwischen dem Granite, den wir als erstes Vorhandenes kennen, und den letzten aufgeschwemmten Gebirgen.
>
> *
>
> Die Hauptschwierigkeit der Geologie beruht auf der Ansicht; darauf nämlich, daß man das Atomistische und Mechanische, welches in gewissen Momenten freilich sich wirksam erweist, solange als möglich zurückdrängt, dem Dynamischen dagegen, einem gesetzmäßig bedingten Entstehen, einem Entwickeln und Umgestalten sein Recht gibt.
>
> *
>
> Wenn man, durch die atomistische Betrachtung, ein bereits Gewordenes hin und her treiben, ablagern und erstarren sieht, so führt die dynamische dagegen in den Moment des Entstehens das lebendige Spiel der Elemente und ihrer Anziehungen ein. In ihr kann sehr vieles noch aus ruhiger Vollstreckung innerer Gesetze hergeleitet werden, was bei jener nur durch einen Aufwand vieler Fluten und äußerer Gewalten begreiflich zu machen ist.
>
> *
>
> Ebenso lehrt uns die dynamische Ansicht partielles Entstehen ohne Schwierigkeit erklären, was bei dem mechanischen und Flutensysteme kaum denkbar ist; sie hält nämlich die ganze Materie für lebens- und verwandlungsfähig, je nachdem es die Bedingungen herbeiführen; sie leugnet ein Kosmisches nicht; sie setzt ein Spiel der Elemente durch die ganze Atmosphäre, mit Anziehungskräften zu dem Festen, wie wir sie jeden Tag, nur modifiziert, gewahr werden; sie sieht ein, daß eine Wechselwirkung zwischen dem Vorhandenen und Entstehenden da sei, durch welche jenes auf dieses, wie dieses auf jenes einwirken könne; sie läßt endlich im bereits Gebildeten noch eine innere Bildung, d.h. eine Sammlung und Anziehung des Ähnlichen und Entsprechenden gelten.

*

Das Unterste, Zugrundeliegende, welches wir auf der Erde gefunden, ist das Granitische. Sein auszeichnender Begriff ist, kein Continens und Contentum, sondern ein vollkommenes Ineinander sein, eine vollkommene Dreieinigkeit seiner Teile zu haben. Sie stehen in ihm gleich und keiner hat ein entschiedenes Übergewicht über den anderen.

*

Gibt das Granitische diesen Charakter auf, so geschieht es dadurch, daß einer seiner Teile ein Übergewicht über die anderen bekommt, seine Weise zu sein zur herrschenden Macht und die übrigen zwingt, nach dieser Weise sich zu gestalten. Der Granit, wo er seinen Charakter aufgibt, hat daher nicht eine, sondern mehrfache Verwandlungsarten". Continues in next footnote.

49 Indem der Granit seinen Charakter aufgibt, tritt also Vielfältigkeit, mehr oder minder stetig, ein. Er hat aufgehört zu herrschen; nun ist eine Anarchie, in welcher jedes zur Herrschaft strebt. In diesem Momente erscheint die Metallformation.

*

Dieses Aufgeben seines Charakters im Granite, diese Metamorphose, kann man als ein Aussichschreiten, ein Überschreiten ansehen. Seine Vielfältigkeit zu bezeichnen, denke man es unter dem Bilde einer Kugel, welche, sowie man aus ihrem Mittelpunkte tritt, Radien nach allen Enden zuläßt.

*

Sehen wir in ihm, daß die Elemente, welche den Granit erzeugen, unter Umständen, zu einem anderen, als er selber ist, können determiniert werden, so ist eine rückschreitende Determination, ein Wiederbestimmt werden des anderen zum Granite gleichfalls nicht völlig undenkbar.

*

Das erste Überschreiten des Granites geschieht in denjenigen Zustand, in welchem hauptsächlich Metallbildung eintritt. Wiewohl Ton und Kiesel in ihm vorherrschend ist, tritt doch auch bereits mannigfach gesäuerter Kalk in ihm ein.

*

Aus diesem Überschreiten geschieht ein zweites. In diesem scheint eine schwankendere Bildung als in dem ersten stattzuhaben. Was eben erst fixiert worden, wird durch neu eintretende Ursachen aufgehoben, verzehrt, gestört, neu hergestellt. Man sucht sich meist mechanisch zu erklären, aber es gehört noch dem Dynamischen an; es ist keine Zerreißung und Ansetzung von außen, sondern ein inneres Aufgelöst- und Neugebunden werden.

*

Doch treten mit und in ihm Erscheinungen ein, bei welchen die Reibung und Spülung der Fluten nicht zu verneinen steht, nur seltener, als man glaubt, und manches, was auf den ersten Blick von Flutungen vorzurücken scheint, wird bei genauer Untersuchung besser von Verwitterung hergeleitet.

*

Bei dem ersten Überschreiten des Granites, also dem ersten erscheinenden Vielfältigen, tritt Kristallisation, also das erste gelungene Individualisieren der Natur ein. Bei diesem Letzten zeigen sich organische Gebilde sowohl großer Farrenkräuter als der Korallen.

*

Dicht am Granite kommt die Grauwacke vor, ein Name für viele Bildungen. Sie trägt große Metaller in sich.

*

Dicht am Granite liegt auch der Porphyr. Dieser hat alle Bestandteile des Granites, aber night in gleicher Herrschaft. Man unterscheidet in ihm das continens von dem contento deutlich. (LA I 1 pp. 379–381)

50 Da dem Menschen nur solche Wirkungen in die Augen fallen, welche durch große Bewegungen und Gewaltsamkeit der Kräfte entstehen, so ist er jederzeit geneigt zu glauben daß die Natur heftige Mittel gebraucht um große Dinge hervorzubringen ob er

sich gleich täglich an derselben eines andern belehren könnte. So haben uns die Poeten ein streitendes uneinig tobendes Chaos vorgebildet. (LA I 11, p. 10)

51 "Mein Geist hat keine Flügel, um sich in jene Uranfänge hervorzuschwingen. Ich stehe auf dem Granit fest, und frage ihn ob er uns einigen Anlaß geben wolle zu denken wie die Masse woraus er enstanden beschaffen gewesen" (LA I 11, p. 10).

52 Many of the terms Goethe used to described granite tors may have been in use at the time. Goethe refers to these as "common mining terms" ["gewöhnlichen bergmännischen Ausdrucke"] (LA I 1, p. 101). But given that one term (Riegel) is denoted with a "?", Goethe may have not been sure which term to use. For example, the term "Kluft" can refer to "fractures" or "joints" (Meschede et al. 2020, p. 188). I have used the term "joint" as it is used to describe the fracturing in granite tors. Riedel et al. (2012, p. 94) refer to joint systems in regional deformation, namely, transverse, diagonal and longitudinal joints ("Q-, S-, L-Klüftung"; see Meschede et al. 2020, p. 188). Given Goethe is referring to jointing (mostly) in massive igneous rocks, which result in the "release 'stored' [of] stress" (Park 1993, p. 6). Rather than using the terms "transverse, diagonal and, longitudinal joints", I have instead translated Goethe's terms from the *Geologisches Wörterbuch (13. Auflage)* and used a synonym to better communicate the terms: Horizontal for "Seam" (*Flöz*; Meschede et al. 2020, p. 108) and Vertical for "Vein" or "Dyke" ("Gang"; Meschede et al. 2020, p. 115). In any case, Goethe uses the terms "waagerechte" (horizontal) and "senkrechten" (vertical) to describe joints further in his *Jointing of Granite*.

53 Massive rock (Massengestein) refers to unstratified rock, such as granite, which is homogeneous.

54 Rock banks (Bänke) are unconsolidated rock matter, such as boulders.

55 "Platten", literally "plates", has a totally different connotation in modern geology (i.e., plate tectonics). Goethe may have been very literally referring to the plate-like structures in granite tors.

56

Wände
  deren Seiten
    Kanten.
Hauptklüfte
Hauptflözklüfte
Hauptgangklüfte
Zwischenklüfte
Zwischenflözklüfte
Zwischengangklüfte.
Massen
  Hauptmassen
  Zwischenmassen
Risse
Ritzen
Schmarren.
Bänke
Platten
Pfeiler
Blöche
? Riegel
Knoten
Keile, spitze,
    abgestumpfte. (LA I 1 pp. 99–100)

57 Die größten Abteilung des Granits, auf die wir zuerst unsre Beobachtung richten, laufen mit der Richtung des Gebirges fort und trennen das Gestein von oben hinab in Massen, die gegen ihre Länge und Höhe schmal sind und deren Seiten parallel miteinander fortlaufen. Wir wollen sie Wände nennen. (LA I 1, p. 100)

58 "Die erste Unterabteilung geschieht durch waagerechte Klüfte, welche von senkrechten durchkreuzt werden; es entstehen daraus große parallelpipedische oder auch, wenn die Klüfte schiefe Richtungen annehmen, rhombische Massen".

59 Goethe uses the term "Ablösung", a term not used in geology (to my knowledge). Synonyms in the *Deutsches Wörterbuch von Jacob Grimm und Wilhelm Grimm* (http://dwb.uni-trier.de/de/), which refer to mining, include "scheiden", that is, parting or dividing, which may be interpreted geologically as "fracturing".

60 Die Granitgebirge bestehen nicht aus ganzen, ungetrennten Massen, vielmehr sind solche in mannigfaltigem Sinne gespalten.

> Die ersten Hauptspaltungen derselben teilen das Gebirge der Länge nach oben hinab. Daraus entstehen Wände von ansehnlicher Breite. Die mittelste Wand steht meist senkrecht, die folgenden zum Teil auch, zum Teil neigen sie sich nach der Mitte zu, und ihre Seiten sind untereinander und mit der Richtung des Gebirges gleichlaufend. Diese Wände werden wieder von andren Ablösungen durchkreuzt. Zuerst bemerken wir einige, die horizontal-Linie entfernen, die ich Flözklüfte nach dem gewöhnlichen bergmännischen Ausdruck zu nennen Vorgänger habe, sodann andre, die mehr oder weniger senkrecht die ersten durchkreuzen und Gangklüfte genannt werden dürfen. Dadurch entstehen Abteilungen, die große parallelepipedische oder rhombische Massen bilden, allein auch diese sind nicht ganz, sondern sie werden wieder durch Unterabteilungen getrennt; deswegen sei mir erlaubt, jene ersten genannten Klüfte Hauptflöz- und Haupgangklüfte, jene Massen Hauptmassen zu nennen.
>
> *(LA I 1, p. 101)*

61 The term "Klüfte" appears later on 21 April 1818 in a ditty dedicated to the birth of his grandson Walther Wolfgang von Goethe (1818–1885). In the poem "Lullaby of the young mineralogist" [Wieglend dem jungen Mineralogen], Goethe recites,

> Peer into the joints [Klüfte] of the mountain,
> As rock quietly develops out of rocks
> [Schau in die Klüfte des Berges hinein,
> Ruhig entwicklet sich Stein ais Gestein].
>
> *(LA I 2 p. 114)*

62 Die Unterabteilungen sind nicht so einfach, noch so regelmäßig und werden ohne Gegenwart des Felsens, ohne Modell oder Zeichnung nicht verstanden werden. Bald laufen die Klüfte dieser Zwischenmassen mit den Hauptklüften parallel, bald weichen sie von ihnen ab. Bald teilen sie solche horizontal in gleiche Bänke oder perpendikulär in gleiche Blöcke, bald trennt sich eine Abteilung in mehrere Langen, indessen die darunter stehende in einen Block verbunden ist, der sich gleichsam wie ein Riegel vorschiebt; bald läuft ein Riß in die nächste Masse hinüber, bald verliert er sich ganz. Die erste Tafel zeigt drei solche Wände, die auf dem Gipfel eines Gebirges stehengeblieben, mit ihren Trennungen in Umriß und beigeschriebenen Termen. (LA I 1, pp. 1001–1102)

63 "Risse und Spaltungen durch Kristallisation, nicht durch Erkältung" (LA I 1, p. 96).

64 Many of the tors either discussed and/or drawn by Goethe are shown in colour photographs in Riedel et al. (2012).

65 An allegory for Goethe's jointing [Zerklüftung] may be found in *Wilhelm Meister's Journeyman Years*, in a conversation between a geologist character, aptly named, Montan

(formerly Jarno in *Wilhelm Meister's Apprenticeship*) and an urchin called Fitz. In the story Fitz shows Montan a mineral that he found nearby. Montan identifies as chiastolite (Al_2SiO_2) [Kreuzstein] from Santiago de Compostela in northwest Spain. Fitz confesses where he got the rock,

> In the ruins of the church at St. Joseph is an altar, also in ruins. Beneath the stones on top, which have all split apart, I discovered a layer of this stone, which served as a base for the other, and I broke off as much as I could. If all the upper stones were moved pout of the way, there would certainly be much more.
>
> *(Goethe 1989, p. 117)*

Fitz is clearly lying as the stone would have been brought to Germany (as it was then) as a souvenir by a pilgrim returning from Spain (someone who the urchin may have pickpocketed). Montan nonetheless paid him his due, a piece of gold [Goldstück].

66 Gestaltung unorganischer Körper

> In flachen Gruben oder Gefäßen erweichter Lehm spaltet sich beim Eintrocknen in fünf- und vierseitige Tafeln.
> Alle Gebirgsarten vom ältesten Granit bis zur letzten Flözschicht spalten sich in gewisse Formen, die mehr oder wenig rhombisch miteinander Ähnlichkeit haben.
> Ziegelsteine, einem allzuheftigen Feuer ausgesetzt, trennen sich in säulenförmige Bildungen.
> Trennung der Masse zu Gestalten.
> Gestaltung der Masse in sich, kristallinisch.
> Kristallisation in Freiheit.
> Abgesonderte Kristallisation innerhalb der Masse.
> Das Porphyrartige.
> Neigung der Trappformation zur Säulengestalt.
> Neigung einer jeden einfachen Gebirgsart zu regelmäßigerer Gestalt.
> Sie kommt nicht immer zur Erscheinung.

67 Also see Miller in Goethe (1995, p. 332, footnote 3) and von Engelhardt (2003, pp. 358–361).

68 Goethe refers to this scene in a second draft of *Faust II* in 17 December 1826, titled *Helena, an interlude to Faust* [*Helena, Zwischenspiel zu Faust*],

> The natural philosophers, who wouldn't miss out on such a opportunity, Thales and Anaxagoras get into an intense argument about the phenomenon, the former ascribing everything to water as well as to moisture, the latter seeing melted and melting masses everywhere [...] both cite Homer, and each calls past and present as witnesses.
>
> [Naturphilosophen, die bei dieser Gelegenheit auch nicht ausbleiben konnten, Thales und Anaxagoras geraten über das Phänomen heftig in Streit, jener dem Wasser wie dem Feuchten alles zuschreibend, dieser überall geschmolzene, schmelzende Massen erblickend, perorieren ihre Solos zu dem übrigen Chor-Gesause, beide führen den Homer an, und jeder ruft Vergangenheit und Gegenwart zu Zeugen].
>
> *(LA I 2, p. 403)*

69

> Kaum wendet der edle Werner den Rücken,
> Zerstört man das Poseidaonische Reich;
> Wenn alle sich vor Hephästos bücken,
> Ich kann es nicht sogleich;

Ich weiß nur in der Folge zu schätzen.
Schon hab ich manches Credo verpaßt;
Mir sind sie alle gleich verhaßt,
Neue Götter und Götzen.

(LA I 2, p. 176; also in Semper 1914, p. 136)

70 *Historical symbols, concerning the genesis of basalt, presented for the argument of parties* [*Historische Symbola, die Basalt-Genese betreffend, zur Einigung der Parteien dargeboten*] (Nose, 1820). The essay titled *Karl Wilhelm Nose* (LA I 8, pp. 157–164; also in LA I 2, pp. 163–171) starts off with a critique, then cites passages from Nose (1820), leaving out original citations given as footnotes in the original. Also Goethe truncates paragraphs possibly in order to emphasise certain points.

71 "Warum sollte man denn dem Basalt, wie er auch erstanden sein möge, ungleiche Zeitepochen seiner Bildung ablegen?" (LA I 2 p, 165).

72 "Wenn der Basalt vulkanisch sein sollte, so mußten Grundlagen gesucht werden, woraus er entstehen können" (LA I 2, p. 166).

73 "Der historische Fingerzeig deutet genugsam auf das Bedürfnis, auf die Notwendigkeit, für jedes vulkanische Produkt ein ursprüngliches Muttergestein aufzusuchen und anzugeben" (LA I 2, p. 167). The unedited original reads, "Der historische Fingerzeig deutet genugsam auf das Bedürfnis – entstell - und verdrängbar freilich durch Künstelei wie alles Natürliche – auf die Notwendigkeit, für jedes vulkanische Produkt ein ursprüngliches Muttergestein aufzusuchen und anzugeben" (Nose 1820, p. 20).

74 See Engelhardt (2003, pp. 299–300).

75 Nehme man ein gutes Beispiel an einem andren mineralogischen Verfahren. Wie sich nämlich in den Steinen und Gebirgen gewisse Pflanzen und Tierformen finden ließen und dies [Wissenschaftlich] szientifisch [sic] zu erwägen war, da betrachtete man die vegetabilischen und animalischen Organismen ebenmäßig als Grundlagen, welche vom Mineralreich aufgenommen, dem selben assimiliert und dadurch verändert, d. i. hier im allgemeinen, versteinert sind. (LA I 2, p. 167)

76 The original reads,

Bekanntlich ist der Basalt mehr oder weniger schmelzbar: der eine mehr, der andere weniger, [d.h.] ie [sic] nach der Mischung seiner Hauptmasse. Unterwirft man ihn nun der Hitze, vom Rot- und Weiß-Glühen an, bis zum Schmelzen und Zerfließen hin: beobachtet man die graduierten Umformungen, welche er dadurch erhält; so ergibt sich ein sicherer Maßstab, wonach jeder Feuer-Effekt auf ihn genau zu bestimmen, und richtig zu beurteilen steht, und der insbesondere dafür geeignet ist, ihn den Phänomenen derjenigen veränderten Basalte anzupassen, welche die Natur in gewissen Gegenden des Erdkreises auffinden läßt.

(Nose 1820, p. 24; also in LA I 2, pp. 168–169)

77 "Die nahe Verwandtschaft der Basalte und Vulkane ist unleugbar [...] Basalte waren also Ausgeburten eines allgemeinen vulkanischen Meeres ..." (LA I 1, pp. 189–191; see also Goethe 1995, p. 135). The existence of "volcanic seas" did exist in the form of Large Igneous Provinces that result in traps, also known as flood basalts (see Marshak 2019, pp. 133–134).

78 "Ob sie echt- order pseudovulkanisch seien, kann die Frage entstehen; aber man neige sich auf welche Seite man will, so wird diesem Falle wegen besonderer Umstände manches problematisch bleiben" (LA I 1, p. 357; also in LA I 8, p. 49).

79 "In der Fassung von 1820 sind hier folgende zwei Abschnitte hinzu- bzw. eingefügt" (LA I 1 p. 358 missing in version of LA I 8).

80 See also Goethe's changing opinion on the volcanic/pseudo-volcanic nature of the Kammerberg in Semper (1914, pp. 117–120), and his discussion with Von Born, Feber and Reuss.

81 The essay is dated "Weimar, den 17. Marz 1824" (LA I 2, p. 331).

82

> Das archetypische Gestein suchen wir in jenen festen Basaltfelsen; wir nehmen an, daß Tonschiefer und Steinkohlen vermischt an dieselben angeschoben worden; diese Gemenge, in der Folge entzündet, hat nicht nur sich selbst verschlackt und ist nach seiner früheren Schichtung auch so verändert übereinander liegen geblieben, sondern die Glut hat auch die anstoßenden Basaltfelsen ergriffen und auf den oberen Teil derselben starken Einfluß gehabt, dahingegen die untersten in ihrer archetypischen Starrheit sich befinden [...] Bedenkt man nun ferner, daß solche Erscheinungen in Böhmen, denen man ihre pyrotypische Eigenschaft nicht absprechen kann, auf dem Ausgehenden der Steinkohlen und Braunkohlenlager sich finden, so wäre man am Ende wohl gar geneigt, diese sämtlichen Phänomene für pseudovulkanische anzusprechen.
>
> *(LA I 2 pp. 330–331)*

83 Das Anschauen gibt uns auf einmal den vollkommenen Begriff von etwas Geleistetem; die Denkkraft, die sich doch auch etwas auf sich einbildet, möchte nicht zurückbleiben, sondern auf ihre Weise zeigen und auslegen, wie es geleistet werden konnte und mußte [...] und so entstehen nach und nach solche Gedankenwesen (enter rationales), denen das große Verdienst bleibt, uns auf das Anschauen zurückzuführen und uns zu größerer Aufmerksamkeit, zu vollkommener Einsicht hinzudrängen. (LA I 1, p. 367)

84 "Man hat diesen Versuch angestellt, um desto deutlicher zu zeigen, daß der in den Schichten des Kammerbergs befindliche mehr oder weniger gerötete Glimmerschiefer durch ein starkes Feuer gegangen" (LA I 1, p. 368).

85

> 7. Glimmerschiefer ohne Quarz, aus Schichten des Kammerbergs. Seine Farbe ist jedoch grau und unverändert.
> 8. Derselbe durchs Porzellanfeuer gegangen, wodurch er rötlich geworden.
> 21. Vollkommene Schlacke.
>
> *(LA I 1, pp. 368–369)*

86 "6. Pyro-technische Versuche durch Natur und Zufall so-genannte pseudo-vulkanische Produkte bei Karlsbad. [...] 7. Pyro-technische Versuche mit Vorsatz durch Töpferfeuer" (LAI 2 p. 151).

87 "Erdbrandsproduckte von der Höhe über Lessau, an der Chaussee nach Schalkenwerth in April/May 1820 [...] 1. Unkenntlichen Gestein", "Gelber Porzellainjaspis mit bläulich, ungleich und rissig [...] Basalt mit Augitkristallen, durchgeglüht. [...] Dergleichen bei dem Dorfe Poppen zwischen Zwodau und Mariaculm" (LA I 2, pp. 141–142).

88 "Pseudovolkanischen Produkte von Hohdorf [Dalovice] mit roten Nummern" (LA I 2 pp. 142–143), "Folge des Quarzgesteins bis zum scheinbaren Konglomerat. Mit großen roten lateinischen Buchstaben" (LA I 2, p. 143), and "Durch das Porzellanfeuer gegangene Bergarten. Mit griechischen kleine roten Buchstaben" (LA I 2, pp. 143–144).

89 In den Anhängen kommt nun die wichtigste Frage vor: inwiefern wir ein Unerforschtes für unerforschlich erklären dürfen, und wie weit es dem Menschen vorwärts zu gehen erlaubt sei, ehe er Ursache habe, vor dem Unbegreiflichen zurückzutreten oder davor stille zu stehen? Unsere Meinung ist: daß es dem Menschen gar wohl gezieme, ein Unerforschliches anzunehmen, daß er dagegen aber seinem Forschen keine Grenze

zu setzen habe; denn wenn auch die Natur gegen den Menschen im Vorteil steht und ihm manches zu verheimlichen scheint, so steht er wieder gegen sie im Vorteil, daß er, wenn auch nicht durch sie durch, doch über sie hinaus denken kann. Wir sind schon weit genug gegen sie vorgedrungen, wenn wir zu den Urphänomenen gelangen, welche wir in ihrer unerforschlichen Herrlichkeit von Angesicht zu Angesicht anschauen und uns sodann wieder rückwärts in die Welt der Erscheinungen wenden, wo das in seiner Einfalt Unbegreifliche sich in tausend und abertausend mannigfaltigen Erscheinungen bei aller Veränderlichkeit unveränderlich offenbart. (LA II 2, p. 171)

90 "Die Wernerische Lehre war eigentlicher Dogmatismus […] aller Dogmatismus der Welt am Ende lästig wird […] Es war nicht zu läugnen daß jene Lehre unerklärte Probleme zurückließ" (WA II 11, p. 307; see also Engelhardt 2003, p. 337).

91

Rhombische Stücke Feldspat
Zugeschärft und in Kristallform übergehend.
Doppel- order Zwillingskristalle.
Sichtbar in vielen Trittplatten an der Häusern her.
Nach aufgelöstem Granit an gewissen Orten merkwürdig und übrigbleibend.
Form derselben.
Doppelkristalle.
Auf- und einander gewaschen.
Im Kreuz entstanden.

(LA I 1, p. 327)

92 "… are called Carlsbad twins after one of their localities, which was where Johann Wolfgang von Goethe first described them" (Hochleitner 2020, p. 306). I'm not sure when the term "Karlsbader Zwillinge" was coined. Goethe uses various terms to describe the twinning. In a letter to August Goethe on 8 August 1814, the elder Goethe still referred to the twins as "Twinned crystals from Carlsbad [Zwillingscrystalle von Karlsbad]" (WA IV, 25, p. 15), yet two years earlier he refers to the twinned crystals as "Carlsbadern" ["… bey Ilmenau sehr schöne und deutliche Zwillingscrystalle gefunden an Gestalt völlig dem bekannten Carlsbadern gleich"] in a letter to Friedrich Wilhelm Heinrich von Trebra (1740–1819) on 7 April 1812 (WA IV, 22, p. 316). I have been unable to find where Goethe uses the term "Karlsbader Zwillinge" in print. If Goethe did coin the term, it was through Weiß that we know about it.

93 *Sammlung zur Kenntnis der Gebirge von und um Karlsbad Angezeigt un Erläutert 1807* (LA I 1, p. 331–346).

94 Große Teile rhombischen Feldspates zeigen sich auffallend in der grobkörnigen Art. Sie deuten sowohl ihrer innern Struktur, als äußern Form nach auf eine Kristallisation, welche sich bald entschiedener ausspricht. Denn es gibt große Massen des Karlsbader Granits, worin man vollkommene Kristalle und zwar von sehr komplizierter Bildung antrifft (5). Es sind Doppelkristalle, welche aus zwei in- und übereinandergreifenden Kristallen zu bestehen scheinen, ohne daß man jedoch den einen ohne den andern einzeln denken könnte. Ihre Form ist durch Beschreibung nicht wohl vor die Einbildungskraft zu bringen, man kann sich solche aber im ganzen als zwei ineinandergefügte rhombische Tafeln vorstellen (6, 7, 8). (LA I 1, p. 331)

95 "Erinnerungsbericht des Mineralogen Chr. S. Weiß August/September 1818" (LA I 1, pp. 118).

96 daß er sie der Kürze halber am liebsten Karlsbader Zwillinge zu nennen pflegte, etwa zu sagen hatte, indem ich das rechts und das links ineinandergewachsen, die einander so ähneln und doch so wenig verwechselbar sind als rechten und linken Arm u. s. f., ihm zu verdeutlichen keine Mühe hatte. Mit einem recht befriedigten und behaglichen Hm! Hm! wiegte er den Kopf vorwärts und rückwärts; dann hab er die Hände ein wenig auf, und indem er

mit ihnen das sogleich zu beschreibende Manöver machte, sagte er: "also die einen sind so! In die anderen so!" Und damit schlug er die Hände zusammen, erst wie gewöhnlich, mit der inneren Seite (den Handtellern) gegeneinander, dann, nachdem er sie umgewendet, beide wieder mit den Handrücken gegeneinander. Das war das ganze Manöver, dessen er bedurfte. Und ich, nicht wenig stutzend im ersten Augenblick, begriff erst bei mehrerem Nachdenken, wie treffend seine bildliche Darstellung war. (LA I 2, pp. 118–119)

97 "Grüßen Sie Herrn Pfarrer Achenbach zum schönsten und lassen die mir in der Ferne gewiesene Stufe Goethit mir nicht verloren sein, die, ob sie gleich den Namen verloren hat, mir doch immer sehr wert sein würde" (Goethe to Cramer, 6 November 1816; WA IV p. 217).

98 "Wohlwollende Männer auf dem Westerwald entdecken ein schönes Mineral und nennen es mir zu Leib and Ehren Goethit; denen Herr Cramer und Achenbach bin ich dafür noch vielen Dank schuldig" (LA I 9, 103; LA I 2, p. 171).

99 "Dieser dunkelrubinrothen, und in zusammengehänste dreiseitige Tafeln crystallisirten Abänderung des Eisenglimmers, welche auf der Eisenzeche zu Siegen sehr sparsam bricht […] hat man daselbst zu Ehren unseres verehrungswürdigsten Präsidenten, des Herrn Geheimenrath von Göthe obigen Namen beigelegt" (Lenz 1806, p. 46).

100 "Rubinrother Eisenglimmer; *Becher*. (Rubinglimmer). Göthit; *Lenz*" (Ullmann 1814, p. 144, original emphasis).

101

1. Braun Eisenstein.

a. Rubinglimmer.
Pyrrosiderit, Göthit, krystallisirter faseriger Braun-Eisenstein, Fer pourpré …

(von Leonhard 1826, p. 231, original emphasis)

102 … denen Herrn Cramer und Achenbach bin ich dafür noch vielen Dank schuldig, obgleich diese Benennung auch schell aus der Oryktognosie verschwand. Es hieß auch Rubinglimmer, gegenwärtig kennt man es unter der Bezeichnung Pyrrhosiderit. Mir war es genug, daß bei einem so schönen Naturprodukt man auch nur einen Augenblick an mich gedacht hatte. (LA I 9, pp. 103–104; LA I 2, pp. 171–172)

103 Another polymorph is Aluminium Goethite (Fe, Al)OOH [Alumogeothit].

104 Goethe never uses the term *Eiszeit* in his work, rather the term was coined by Schimper in 1837 (Günther in Magnus 1949, pp. xvii–xviii). The term was introduced as a title in the *Leopoldina Ausgabe* (LA I 2, p. 390) by the editor Günther Schmidt to denote Goethe's writings on the topic.

105 "Eis spielt eine größere Rolle, als man denkt" (LAI 2, p. 390).

106 Also see LA I 11, pp. 309–310.

107 Umherliegende Granite (WA II 10, p. 95; translated Cameron 1965, p. 751).

108 Kälte (WA II 10, p. 95; translated Cameron 1965, p. 752).

109 Erratische Blöcke (LA I 2, pp. 377–378; translated Cameron 1965, p. 752).

110 "… für das Musterstück des athmophärischen Steins" (Goethe to Blumenbach 17.9.1803, WA IV 16, p. 303). Goethe noted in 1806–1807 that he had "found the most credible report of a meteorite in the Thuringian Chronicle" [die glaubwürdigste Nachricht von einem Meteorstein in der Thüringer Chronik fand] (LA I 1, p. 304). Goethe may be referring to the Meuselbach meteorite, which fell in Meuselbach, Thuringia in 1897.

111 Eine ganz eigne Empfindung ist es, wenn ich nun gerade beide gemeldete Werke neben einander lege und in beiden unbegreifliche Naturprodukte vor mir sehe. Vegetablien, von denen nicht abzusehen ist, wie sie nahe den fünfzigsten Grade unserer Breite wachsen könnten; Mineralien, die aus heiterer Luft, wo man dergleichen nicht vermuten, sich auf einmal konzentrierend herunterstürzen. Das lassen wir denn alles gern geschehen, sobald der aufmerksame Beobachter uns dergleichen Erscheinungen im großen

allgemeinen Zusammenhang vorlegen und anschaulich machen will, wodurch denn ganz allein der Geist gegen solche ungeheure Gegenstände in Freiheit geraten kann. (Goethe to Schreibers, 7.1.1821, WA IV, 34, p. 73)

112 Auch kann ich nicht läugnen, daß ich weiterhin selbst weder gern von körnigem Weißstein, noch von neuerem Granit und Weissstein reden möchte. Wäre es mir daher erlaubt, einen Zwischenweg einzuschlagen, und einen Namen in Vorschlag zu bringen, der auf beide Abänderungen passte, so möchte es wohl der Name Granulit sein. (Weiß 1803, p. 348)

113 In which orthopyroxene appears in metabasite, that is, metamorphosed mafic rocks, such as gabbros, basalts and dolerites.

114 Specimen No. 2927 from Catalogue A (A4 Suite from the Harz)

> 22. Merkwürdiges Gestein aus dem Budetal der Suseburg schief gegenüber, das ich mir weder zum Granit, noch zum Porphyr zu rechnen getraue, und für welches ich den Namen Granulat, wegen der in solchen befindlichen Quarzkörnern, in Vorschlag bringe. Es ist äußerst fest.
>
> *(LA I 1, p. 84)*

The rock was collected by Goethe in 1784.

115 "Der von Goethe für ein *merkwürdiges Gestein* vorgeschlagene Name entspricht nicht de heutigen, auf Weiß zurückgehenden Benennung eines hellen, geschieferten Gesteins aus Quarz und Feldspat mit Granat sowie Disthen und/oder Pyroxen" (Kuhn and von Engelhardt in LA II 7, p. 593, original emphasis).

116 No. 2927 in "Katalog A" (Prescher 1978, p. 166).

117 *Germany geognostically geologically represented, with Maps and Cross-sections, which comprise a geognostic Atlas. A Periodical* [*Teutschland geognostisch-geologisch dargestellt, mit Charten und Durchschnittszeichnungen, welche einen geognostischen Atlas bilden. Eine Zeitschrift*] (Keferstein 1821).

118 "… with Verehrung und Dankbarkeit gewidmet" (Keferstein 1821, p. iii).

119 Bei der Illumination der Charten kam es nicht allein darauf an, durch verschiedene Farben verschiedene Formationen anzuzeigen, sondern auch darauf, die Farben so zu wählen, daß sie zweckmäßig verteilt sind und angenehm in's Auge fallen; hierbei war die Beihülfe eines großen Kenners der Farben von hoher Wichtigkeit. Herr Geheimde-Rat von Goethe war so gütig, sich nicht allein für dieses Werk im Allgemeinen zu interessieren, sondern auch eine Farbentafel anzugeben, welche der Illumination des ganzen geognostischen Atlasses zu Grunde liegen wird, und ich ergreife diese Gelegenheit, meinen innigsten Dank dafür hier wiederholen zu können. (Keferstein 1821, p. x)

120 "Wird nun der intendierte geognostische Atlas auf solche Weise durchgeführt so wäre zu wünschen, daß die Freunde dieser Wissenschaft sich vereinigten und dieselben Farben zur Bezeichnung eben desselben Gesteins anwendete, woraus eine schnellere Übersicht hervorträte und manche Bequemlichkeit entstünde" (LA I 2, p. 192).

121 Der Hauptformation, welche Granit, Gneis, Glimmerschiefer mit allen Abweichungen und Einlagerungen enthält, erteile man die Karminfarbe, das reinste schönste Rot; dem unmittelbar anstoßenden Schiefer gab man das harmonierende reine Grün; drauf dem Alpenkalk das Violette, auch dem Roten verwandt, dem Grünen nicht widerstrebend. (LA I 2, p. 191)

122 "Sinnlich-Stittliche Wirkung der Farbe" (LA I 4, p. 224).

123 The earliest known colourised geological map, titled "Mountains of Bekhen", is found on the Turin papyrus from either the 19th or 20th Dynasty, New Kingdom Egypt (ca. 1150 BCE; Romer 1989, p. 130). In the map, granite, as well as volcanics, are denoted in pink, which is the same colour used in Keferstein's *General Charte von Teutschland.* The similarities are mostly likely due to the colour of the rock (pink in both cases) being reflected in the colour classification within both maps. Granites are usually pink due to the potassium (alkali) feldspar mineral in the rock, namely orthoclase.

124 See Piper (2010) for a fascinating account of the impact geology (granite) and cartography had on Goethe's later novels. Also see Steer (1979) on another take on the impact Goethe's science had on the structure of *Wilhelm Meister's Journeyman Years*. Note that Steer (1979) does not mention cartography, geology or granite.

125 Goethe, I believe, is referring to paragneiss, a metamorphic rock derived from sedimentary rocks, as opposed to an orthogneiss, which may be derived from granite.

126 Goethe acknowledged that observation in geology was influenced by the observer's school of thought: "Such a poem [King Coal's Levee (Scafe 1818)]," said Goethe,

> is quite calculated to amuse people of the world; while at the same time it diffuses a quantity of useful information, which no one ought properly to be without. A taste for science is thus excited amongst the higher circles; and no one knows how much good may ultimately result from such an entertaining half-joke. Many a clever person may be induced to make observations himself, within his own immediate sphere. And such individual observations, drawn from the natural objects with which we are in contact, are often the more valuable, the less the observer professionally belongs to the particular department of science.

"You appear, then, to intimate", returned I, "that the more one knows, the worse one observes".

"Certainly", said Goethe,

> when the knowledge which is handed down is combined with errors. As soon as any one belongs to a certain narrow creed in science, every unprejudiced and true perception is gone. The decided Vulcanist always sees through the spectacles of a Vulcanist and every Neptunist, and every professor of the newest elevation-theory, through his own. The contemplation of the world, with all these theorists, who are devoted to an exclusive tendency, has lost its innocence, and the objects no longer appear in their natural purity. If these learned men, then, give an account of their observations, we obtain, notwithstanding their love of truth as individuals, no actual truth with reference to the objects themselves but we always receive these objects with the taste of a strong, subjective mixture
>
> *(Goethe to Eckermann, 18 May 1824).*

REFERENCES

Appel TA (1987) *The Cuvier-Geoffroy debate: French biology in the decades before Darwin*. Oxford University Press, Oxford.

Baldridge WS (1984) The geological writings of Goethe: Despite his keen powers of observation, Goethe's ideas on geology reflected the biases of his time. *American Scientist*, 72: 163–167.

Bakewell R (1815) *An introduction to geology*. J. Harding, London.

Böhme H (2016) *Natur un Figur: Goethe im kontext*. Wilhelm Kink, Paderborn.

Borghini A, Ferrero S, O'Brien PJ, Laurent O, Günter C, Ziemann MA (2019) Cryptic metasomatic agent measured in situ in Variscan mantle rocks: Melt inclusions in garnet of eclogite, Granulitgebirge, Germany. *Journal of Metamorphic Geology*. https://doi.org/10.1111/jmg.12519

Boyle N (1991) *Goethe: The poet and the age*. Oxford University Press, Oxford.

Brochant AJM (1803) *Traité Élémentaire De Minéralogie* (Vol. 2). Chez Villier, Paris.

Caisley JS (2020) The (im)materiality of Goethe's geology. PhD Thesis, University of Cambridge, Cambridge.

Cameron D (1965) Early discoverers XXII Goethe – discoverer of the Ice Age. *Journal of Glaciology*, 5: 751–754. https://doi.org/10.3189/S0022143000018761

Coutinho J, Kräuter H, Sassi F, Schmid R, Sen S (2007) Amphibolite and Granulite. In *Metamorphic Rocks: A classification and glossary of terms* (Eds. D Fettes, J Desmond) Cambridge University Press, Cambridge, pp. 51–57.

de Saussure B-H (1779) *Voyages dans les Alpes*. Chez Samuel Fauche, Neuchatel.

Engelhardt VW (2003) *Goethe im Gespräch mit die Erde*. Verlag Böhlaus Nachfolger Weimar, Weimar.

Fink KJ (1991) *Goethe's history of science*. Cambridge University Press, Cambridge.

Goethe JW (1989) *Conversations of German refugees and Wilhelm Meister's Journeyman Years or The Renunciants* (Ed. JK Brown). Suhrkamp Press, New York.

Goethe JW (1995) *Scientific studies* (Vol. 12) (Ed. D Miller). Goethe: The Collected Works. Princeton, New Jersey.

Haeckel E (1866) *Generelle Morphologie der Organismen: allgemeine Grundzüge der organischen Formen-Wissenschaft, mechanisch begründet durch die von Charles Darwin reformierte Deszendenz-Theorie*. G. Reimer, Berlin.

Haeckel E (1876) *The history of creation*. Henry S. King and Co., London.

Hamm EP (1991) Goethe on Granite. PhD Thesis, University of Toronto, Toronto.

Hamm EP (2001) Unpacking Goethe's collections: The public and the private in natural-historical collecting. *The British Journal for the History of Science*, 34: 275–300.

Hennig J (1949) Goethe's interest in British mineralogy. *Mineralogical Magazine and Journal of the Mineralogical Society*, 28: 534–546. https://doi.org/10.1180/minmag.1949.028.205.02

Hochleitner R (2020). *The minerals encyclopedia*. Firefly Books, Stuttgart.

Horn S, Kreher-Hartmann B, Heide K (2001) Melting experiments and field work on Komorní Hùrka volcano, Bohemia, by Johann Wolfgang von Goethe. *Journal of Geodynamics*, 32: 77–97.

Jameson R (1808) *System of mineralogy*. Archibald Constable and Co., Edinburgh.

Keferstein C (1821) *Teutschland geognostisch-geologisch dargestellt, mit Charten und Durchschnittszeichnungen, welche einen geognostischen Atlas bilden*. Verlage des Landes-Industrie-Comptoir, Weimar.

Lacroix A (1901) *Minéralogie de la. France et de ses Colonies* (vol. 3). Ch. Béranger, Paris.

Lenz JG (1806) *Tabellen über das gesammte Mineralreich*. Göpferdts, Jena.

Leonard K et al. (1817) *Propaedeutik der Mineralogie, Joh. Christ*. Hermannschen Buchhandlung, Frankfurt am Main.

Leonhard C (1826) *Handbuch der Oryktognose*. JCB Mohr, Heidelberg.

Ludwig CF (1804) *Handbuch der Mineralogie nach A. G. Werner*. Siegfried Lebrecht Crusius, Jena.

Magnus R (1949) *Goethe as a scientist*. Henry Schuman, New York.

Marshak S (2019) *Essentials of geology* (6th edition). W.W. Norton, New York.

Meixner S (2002) Urphänomen (Original/Primordial Phenomenon). *Goethe-Lexicon of Philosophical Concepts*, 3. https://doi.org/10.5195/glpc.2022.46

Meschede M, Murawski H, Meyer M (2020) *Geologisches Wörterbuch* (13th edition). Springer, Berlin.

Nelson G (1978) Ontogeny, phylogeny, paleontology, and the biogenetic law, *Systematic Biology*, 27(3): 324–345. https://doi.org/10.2307/2412883

Niklas KJ, Kutschera U (2017) From Goethe's plant archetype via Haeckel's biogenetic law to plant evo-devo 2016. *Theory in Biosciences*, 136: 49–57. https://doi.org/10.1007/s12064-016-0237-7

Nose KW (1820) *Historische Symbola, die Basalt-Genese betreffend, zur Einigung der Parteien dargeboten*. Eduard Weber, Bonn.

Oldroyd DR (1979) Historicism and the rise of historical geology, part 1. *History of Science*, 17: 191–213. https://doi.org/10.1177/007327537901700303

Ospovat AM (1959) Werner's influence on American geology. *Proceedings of the Oklahoma Academy of Science*, 40: 96–103.

Park RG (1993) *Foundations of structural geology*. Chapman and Hall, New York.
Piper A (2010) Mapping vision: Goethe, cartography, and the novel. In (Eds. J Fisher and B Mennel) *Spatial turns: Space, place, and mobility in German literary and visual culture*. Rodopi, Amsterdam, pp. 27–51.
Prescher H (1978) *Goethes Sammlungen zur Mineralogie, Geologie und Paläontologie: Katalog*. Akademie-Verlag, Berlin.
Reimarus JAH (1802) *Bildung des Erdballes und ins Besondere über das Lehrgebäude des Herrn de Luc*. Carl Ernst Bohn, Hamburg.
Riedel G-R, Klauß J, Feiler H (2012) *Granit: Goethes Urgestein*. Schibri-Verlag, Uckerland, Brandenburg.
Romer J (1989) *Ancient lives*. Phoenix Press, New York.
Scafe J (1818) *King Coal's leave, or geological etiquette: With explanatory notes; and the council of the metals*. J Graham, Alnwick.
Schäfer-Weiss D, Versemann J (2005) The influence of Goethe's Farbenlehre on early geological map colouring: Goethe's contribution to Christian Keferstein's general Charte von Teutschland (1821) *Imago Mundi*, 57(2), 164–184. https://doi.org/10.1080/03085690500094990
Schneider S (2021) Goethes Archivbegriff und Archivpoetik am Beispiel einer Szene aus Faust II. (Eds. K Kastberger, C Neuhuber), *Archive in/aus Literatur: Wechselspiele zweier Medien*. De Gruyter, Berlin, pp. 13–28. https://doi.org/10.1515/9783110742503-002
Schrank FVP (1793) *Anfangsgründe der Bergwerkskunde*. Johann Wilhelm Krull Buchhändler, Ingolstadt.
Schroeder D (1988) Zur Geschichte des Mineralnamens "Goethit". *Zeitschrift für Pflanzenernährung und Bodenkunde*, 151, 137–139. https://doi.org/10.1002/jpln.19881510213
Semper M (1914) *Die geologischen Studien Goethes*. Verlag von Veit and Comp., Leipzig.
Sosman RB (1916) Types of prismatic structure in igneous rocks. *The Journal of Geology*, 24: 215–234. https://doi.org/10.1086/622324
Steer AG (1979) *Goethe's Science in the structure of the Wanderjahre*. The University of Georgia Press, Athens.
Steigerwald J (2002) Goethe's morphology: Urphänomene and aesthetic appraisal. *Journal of the History of Biology*, 35: 291–328. https://doi.org/10.1023/A:1016028812658
Steiner W (1957) Zur Geschichte der geologischen Karte. *Zeitschrift fur angewandte Geologie*, 3: 417–422.
Steiner W (1969) Christian Keferstein. Ein Wegbereiter der regjonalen Geologie Deutschlands. Berichte der Deutschen Gesellschaft fur Geologische Wissenschaften. *Reihe A: Geologie und Paldontologie*, 14: 269–332.
Ullmann JC (1814) *Systematisch-tabellarische Übersicht der mineralogischen-einfachen Fossilen*. Kriegerischen Buchhandlung, Kassel.
von Zittel KA (1899) *Geschichte der Geologie und Paläontologie bis Ende des 19*. Jahrhunderts. Druck und Verlag von R. Oldenbourg, München und Leipzig.
Wagenbreth O (1967) Abraham Gottlob Werner und seine Bedeutung fur die Entwicklung der geologischen Landesaufnahme und des geologischen Kartenwesens. *Zeitschrift fur angewandte Geologie* 13: 372–384.
Watt G (1804) XI. Observations on basalt, and on the transition from the vitreous to the stony texture, which occurs in the gradual refrigeration of melted basalt; with some geological remarks. *Philosophical Transactions of the Royal Society*, 94: 279–314. https://doi.org/10.1098/rstl.1804.0013
Weiß CS (1803) Über die Gebirgsarten des sächsischen Erzgebirges, ein Schreiben an Herrn Geh. *Ober-Bergrath Karsten mit einigen Anmerkungen von dem Letztern Der Gesellschaft Naturforschender Freunde zu Berlin Neue Schriften*, 4: 342–366.

Wells GA (1978) *Goethe and the development of science 1750–1900*. Sijthoff and Noordhoff, Alpha can den Reign, The Netherlands.
Werner AG (1787) *Kurze Klassifikation und Beschreibung der verschiedenen Gebirgsarten*. Waltherischen Hofbuchhandlung, Dresden.
Wilkinson EM (1951) Goethe's conception of form. *Proceedings of the British Academy*, 37: 175–197.

3 Goethe on Colour

Goethe composed several works on colour, including the two-volume *Beiträge zur Optik* [*Contributions to Optics*] (1791–1792) and the three-volume *Zur Farbenlehre* [*Theory of Colours*] (1808–1810). Goethe first proposed his method in *Beiträge*, which was later developed further in *Zur Morphologie* [*On Morphology*] and *Farbenlehre*, as well as in several smaller works. The concept of Anschauung and the Urphenomenon were developed by his learning-by-doing method; namely, by doing his experiments via observation (i.e., Anschauung), the reader would eventually perceive the Urphenomenon.

Goethe's first work, *Beiträge zur Optik*, consists of two volumes, including a set of 27 playing cards that depict images for use in his experiments using a clear glass prism. The work stopped prematurely, as only two further volumes of the 6 planned were written in draft form.

The period between *Beiträge* and *Farbenlehre* Goethe entered his "Transition" period (Matthaei in Goethe 1971) where he concentrated further on developing his method fully. Between 1792 and 1808 Goethe penned several essays and letters, such as *Empirical observation and science* ["*Das reine Phänomen*"] in 1798, and *Symbolism* [*Symbolik*] and *Introduction to the Lectures on Physics* ["*Einleitung zu physikalischen Vorträgen*"] in 1805. It is also important to note that during that time Goethe started writing his *Zur Morphologie* (1810), meaning that his method was cross-pollinating between his work on optics and biological form.

Farbenlehre, composed between 1808 and 1810, published in two volumes in 1810, is divided into three sections, the Didactic, the Polemic and Historical. Goethe's method is mostly discussed in the didactic section. The polemic section is a negative critique of Newton's *Opticks* (Newton 1704), which also contains an array of experiments. Not known to many is that the historical section is possibly the first modern and comprehensive example of a history of science. Along with the polemic, the historical section also contains a critique of Newton.

The primary concern of this chapter is to reveal Goethe's methodology, rather than dwell on his critique of Newton, which is extensively covered in modern literature (see Sepper 1988). Goethe's Polemic section, which was not translated in full until Duck and Petry in Goethe (2016), along with the historical section, was absent from the first translation by Eastlake in Goethe (1840) because it lacked any "scientific value". The historical section remains untranslated.

Goethe considered his *Farbenlehre* to be his greatest contribution. In his conversation with Eckermann, Goethe noted:

> In terms of everything that I have accomplished as a poet [...] I don't have any illusions. There have been excellent poets who lived at the same time as I, and those who were even better before me, and there will be more of them after me. That I am the only one in my century who is right in the difficult science of *Farbenlehre* – that I do attribute to myself and I am conscious in that regard of being superior over many.[1]

DOI: 10.1201/9781003275343-3

Conversely, Goethe's work on colour has been considered one of Goethe's greatest failures, generally because it was misunderstood as an alternative theory to, or an improvement on, Newton's *Opticks*. This incorrect and stereotypic view of Goethe's criticism of Newton is outlined by Dennis Sepper.

> First, the poet, dedicated to the richness of life, objected to mathematical physics' analytical decomposition of natural experience. Second, although one may sympathise with the poet's fears for nature, when the purposes of knowledge are at stake it is necessary to defend the objectivity and truth of science.

Sepper continues,

> It is supposed that the German poet, prepossessed (that is, prejudiced) by the simple, bright appearance of white light, could not accept that it is composed of innumerable colours. Coloured light is darker than white light, so no matter how many crucial experiments Newton might have performed with his prisms and lenses, the poet could not accept that white light is the sum of many darknesses. Goethe was destined by his poetic character to draw a fallacious conclusion from sensation and thus to reject Newton, experimental method, and the application of mathematics to nature. In a word, he was destined to reject modern science.
>
> *(Sepper 2009, p. 261)*

Goethe didn't reject Newtonian mechanics or mathematical analysis, something that is often attributed to Romanticism,[2] but rather argued that looking through a glass prism gives you a different phenomenon than the one Newton had predicted in his theory of differential refrangibility of light rays. Unbelievably the Newtonians never bothered to check, that is to observe, whether the colours they predicted were produced.[3] To understand Goethe's Morphology, we need to circumvent the Newton *contra* Goethe debate and understand the following.

> There is a delicate empiricism which identifies itself with the object [Gegenstand] in the most intimate way and thereby becomes the actual theory [Theorie]. But this increase in intellectual ability belongs to a highly educated age.
>
> The highest would be to understand that everything factual is already theory. The blue of the sky reveals to us the basic law of chromatics. Just don't look for anything behind the phenomena; they themselves are the teaching [Lehre].
>
> When I finally settle from the Urphenomenon, it is only resignation; but there remains a great difference whether I resign myself to the limits of humanity or within the hypothetical constraints of my narrow-minded individuality.[4]

Goethe's *Farbenlehre* is an atheistic appreciation of colour in the same way his *Zur Morphologie* (1817) is an atheistic appreciation of organic form. Unfortunately, many who read Goethe's work on colour do not appreciate that not all science is experimental, where a hypothesis is tested via experimentation, which can be repeated. Unlike experimental science, historical science (e.g., taxonomy, phylogenetics, petrology, mineralogy, physical geography, soil science) is limited to observation and measurement as past processes are unobservable and only proposed via inferences or argumentation (see Wilkins and Ebach 2014). Moreover, historical science is not as hypothesis driven as modern experimental science. Think of when a new organism or mineral is collected, observed, measured and eventually discovered to be a new species. There is little in the way of hypothesis other than "What's that?" Goethe method belongs in the historical sciences, where the mantra isn't hypothesis testing, but rather observe, collect and record. For example, any taxonomist, soil scientist or mineralogist reading Goethe will immediately relate to his method as something they do in their practise of historical science.

A *THEORY* OF COLOUR?

Farbenlehre is often translated as a *theory* of colour. Goethe does use the term *theory*, but not in the modern usage; rather, in the root meaning, *theoria*, "to see", as in Anschauung (see Chapter 1). In this sense the term theory becomes confusing to a modern reader who expects to be used as a noun, namely "an explanation of a phenomenon" or "a statement of one or more laws and principles".[5] On the first translation of *Farbenlehre*, Eastlake (1840) translates the title as *Goethe's Theory of Colours*, but refers to *Farbenlehre* in the Translator's Preface as *Goethe's Doctrine of Colours*.[6] Which is it? An explanation or an activity? Herein lies the crux of Goethe's method. If we read *Farbenlehre* as an explanation, then we get a very different impression of it than if we view it as an activity. But more on this later.

How to best translate *Farbenlehre*? We know that *Farben* refers to "colour". The problem is what "Lehre" refers to particularly in 19th-century German, given languages change over time. In the German Dictionary by Jacob Grimm and Wilhelm Grimm, "Lehre" is defined as

> … the entire content of a lesson, that which is taught in a coherent manner in a branch of knowledge.[7]

In any modern dictionary, this would be translated as "theory" or even as "science".[8] Grimm and Grimm also provide an example for their definition from Goethe's *Farbenlehre*,

> … appears in multiple connections if Newton's teaching [Lehre] was easy to learn, insurmountable difficulties were revealed in its application. Our teaching [Lehre] is perhaps more difficult to grasp, but then everything is done; these colours, which form the foundation of the entire teaching [Lehre].[9]

Lehre in this sense is better translated as a verb: "teaching", "training" or "learning". *Farbenlehre* is meant to be dynamic, to be practised as the reader reads and interacts with the experiments in the book. When you read *Farbenlehre*, Goethe is showing how you can interact with colour and light, such as how light interacts with darkness, how colours mix and how they change our mood. There is no *theory* (i.e., explanation) as such, other than Goethe's definition of *theoria*, to see.[10] A synonym of Lehre is Anschauung,[11] which is probably closer to the intended meaning of the work.

Farbenlehre is to be practised by the reader and in this sense, *Zur Farbenlehre* would more accurately be termed "Toward a [Theory] Practise of Colour". Given that the practise is Anschauung, it would more accurately be translated as "Towards Seeing Colour". Goethe's earlier *Beiträge zur Optik* (1791–1792) is translated as "Contributions to Optics" (see Matthaei in Goethe 1971, p. 6, and below). To avoid any contention, I will use the original German herein for all of Goethe's works on colour, namely, *Beiträge zur Optik* or *Beiträge* and *Farbenlehre*.[12]

BEITRÄGE ZUR OPTIK (1791–1792)

Morphology features little in Goethe's *Beiträge*, except as an early albeit unstructured version of learning-by-doing, namely a hands-on and task-driven exercise that explores colour by looking at colour playing cards (and plates) through a glass prism. Goethe is in conversation with the reader and together you conduct these colour exercises by holding the glass prism (Figure 3.1) to your eyes and observing the 27 colour cards or plates (Figure 3.2). As you observe, Goethe describes what you see and how you can manipulate the cards or prism to get varying effects. Essentially Goethe is teaching you an activity, namely, Anschauung, and in doing so shows you how to form the Urphenomenon. By 1792, these terms have not entered the Goethean

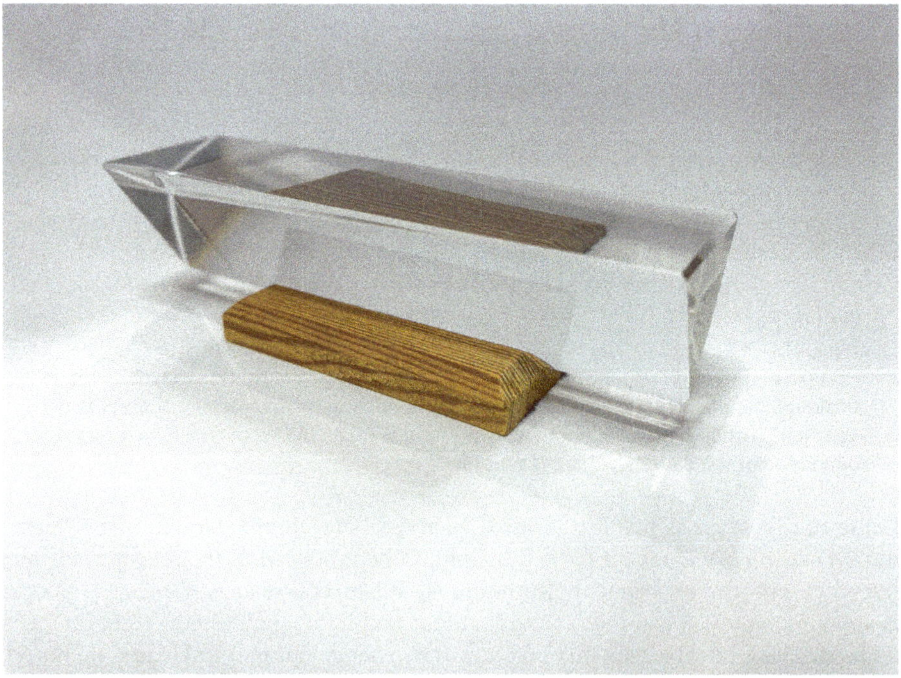

FIGURE 3.1 A glass prism 10 cm in length used by the author.

(Photo by the author, 2026)

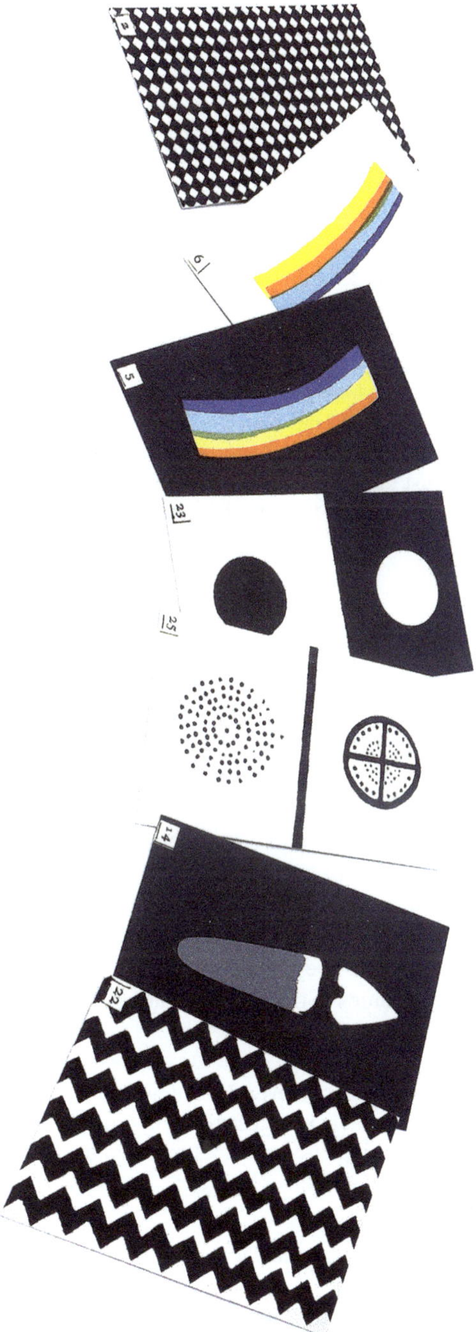

FIGURE 3.2 A selection of the 27 playing cards accompanying *Beiträge zur Optik*. **(Photo by the author, 2026)**

Lexicon (see Chapter 1), but the actions are clearly there. Goethe is showing you how observation is an empirical science.[13]

> My intention is to collect all the empirical evidence in this area, do every experiment myself, and develop the experiments in their most manifold variations so that they become easy to reproduce and more accessible. I will then attempt to establish the axioms in which the empirical evidence of a higher nature can be expressed and see if these can be subsumed under still higher principles.[14]

Goethe's *Beiträge* takes the reader on a journey into the realm of phenomena.

> We believe the blue character of the air, since we see the sky as blue, simply as a result of initial awareness when confronted by a large volume of air. We also interpret the blueness of mountains in this manner. Yet if we pay closer attention to this, it becomes an insufficient explanation. For, if this were correct, the forest mountains would appear as the darkest blue, since the largest volume of air is between us and the forest mountains. But we observe just the reverse, for the mountains appearing a beautiful, intense blue only at a certain distance, becoming lighter blue as they recede and ultimately disappearing in a white haze.
>
> *(§6; Goethe 1971, pp. 14–15)*

I too have recently had this experience in the area where I live, in the Blue Mountains, New South Wales, Australia. I saw the same phenomenon from the lookout at Wentworth Falls in the Blue Mountains (Figure 3.3).

Goethe's observation is free from any speculative theory (i.e., models) or inferences. No assumed unobservable objects, particles or waves are needed to explain what we can simply observe on our own. Matthaei (in Goethe 1971) notes that the view "disappearing in a white haze" has the hallmarks of Goethe's pure or Urphenomenon. Our Anschauung is listed to what we can experience, which in turn limits the extent of our Urphenomenon. In Figure 3.3 the colour blue nearer in view is a more intense blue, which over greater distances turns into a light blue then into a whiteish colour. In our mind's eye, we are limited to that experience of the Urphenomena. Goethe continues to discuss the concept of the Urphenomena (without mentioning the term) in §8 and §9.

> Since ancient times thoughtful people have been drawn to these beautiful or, under certain circumstances, uncomfortable manifestations, in order to view them more closely, to reproduce them under varying circumstances by ingenious experiments, to close the gap as to their cause and relationships. The history of optics shows us how slow progress has been.
>
> *(§9; replace translation that is found here: Goethe 1971, p. 15)*

The "uncomfortable manifestations" Goethe refers to are the atmospheric phenomena, such as the blueness of mountains or the rainbows that occur during a thunderstorm. Goethe is showing us how phenomena are steadfast in appearance. The blueness fades the further we look into the distance and the rainbow always has the same series of colours that petter out at each end, after which the whole phenomenon

FIGURE 3.3 Lookout over a frozen Wentworth Falls, Blue Mountains, NSW, Australia. Note how the blue fades into the distance.

(Photo with permission Julie Martin, 2018)

fades away. The emergence and dematerialisation of these phenomena, such as the blueness and the rainbow, show us the limitations of our own experiences and the Urphenomenon. We are unable to conjure up new colours or make objects in the distance more blue.

These examples show where Goethe differs to the physics of his day. Physicists were interested in inferring a natural process that sought to explain the phenomena rather than the experience of the phenomena. That desire to explain, rather than to observe, phenomena led Goethe straight to Newton, whom he referred to as "a thoughtful man" whose own experiments in *Opticks* were "complicated and hard to repeat". Goethe continued,

> Further, the theory is abstract, and its application cannot be judged without the most profound insight into higher mathematics.
>
> *(§13; Goethe 1971, p. 16)*

There are two parts to this statement, which is open to misinterpretation. Firstly, Goethe saw theory as practise, not as an abstraction, hence "abstract theory". Newton's method was not in any way an activity in that it featured Anschauung or developed an Urphenomenon. It was just reporting colours and predicting what colours would form based on mathematics. Secondly, Goethe was not decrying mathematics or the mechanisation of natural history. Goethe saw the importance of mathematics. He just didn't see its use in observing phenomena. In fact, it simply got in the way, as though the experience got in the way of the maths. Goethe's later outrage in his Polemical Part in *Farbenlehre* goes into greater detail about how the cart (mathematics) is placed in front of the proverbial horse (experience of colour, Anschauung and the Urphenomenon). Goethe states this unequivocally,

> These difficulties would have discouraged me had I not reflected that pure experiences [reine Erfarungen] should lie at the root of all physical science; that these experiences can be arranged in sequence without taking into account any other considerations; that a theory can be judged worthy only when all experiences are brought under one roof and assist their subsequent application [...] It is with this conviction that I decided to investigate that physical portion of the principles of light and colour without presuming any other consideration even if there were many doubtful issues and new data to be discovered.
>
> *(§14; Goethe 1971, p. 16)*

The pure experiences Goethe refers to are interpreted by Matthaei (in Goethe 1971, p. 16) as "secured observations whose conditions are known". My interpretation would be Anschauung, even though Goethe didn't use the term for such experiences until 1795. Goethe's critique of Newton's method is more or less a critique of physical science at the time. Yet, this critique drove him to note that it is his "duty to undertake and accurately repeat the renowned experiments, to analyse, to compare and to classify, thus enabling me to conduct additional experiments". Goethe wanted to

> make my country more aware of this science, at least more so than it has been up to now. Consequently, I have gone to great pains to enable anyone to undertake the practical experiences under discussion easily. At the end of this section I will specifically discuss the utility of these small diagrams published simultaneously with them.
>
> *(§15; Goethe 1971, p. 16)*

Those small diagrams are his set of playing cards, which I will discuss below.

Before we continue our conversation with Goethe, it is important to understand why he was so passionate about observation. While navigating the maze of Venetian alleyways (the *calli* and *callettes*) during his Italian journey, Goethe was adamant about honing-in his senses when practising his "method".

> Finally one does disentangle oneself, but it is an incredible maze, and my method, which is to acquaint myself with it directly through my senses, is the best.
>
> *(30 September 1786; Goethe 1989, p. 60)*

By 1786 Goethe had expressed no desire to study colour. Rather his interests lay elsewhere in geology and comparative anatomy. Imagine his surprise that his method of observation, an activity, has been usurped by a passive abstract theory based on mathematics! Whether this explains Goethe's outburst in his *Beiträge* is of course hearsay. I only mention it because he already had a method in mind before he studied colour. The origins of Goethe's method lay not in the streets of Venice, but rather a passion that most likely was one of the reasons that drove the polymath to Italy in the first place, namely, art.

Before Goethe started experimenting with colour, his *Beiträge* introduces those engaged in colour theory, namely painters.

> The creative artist could gain but little advantage from a theory whereby the optical scientist by his negative efforts merely explains overall occurring phenomena.
>
> *(§19; Goethe 1971, p. 18)*

What painter would gain any experience of colour simply through abstract theory? But what method did the creative artist use? Goethe called out what many had not, namely that

> … these artists could scarcely communicate the rules upon which they operate.
>
> *(§19; Goethe 1971, p. 18)*

In other words, what were the artists doing? We can see a painter compose a painting, but what rules are being employed? What was their *method*?

While Goethe did not emphasise the importance of this statement (he presumed each artist had their own method and lay the matter to rest), he however did note that artists did discover the law of light and shade, a concept later to become known as *Polarity* (see Matthaei in Goethe 1971, p. 18), an idea that was later to become the Urphenomenon (see Miller in Goethe 1995, p. 333). While much of Goethe's Morphology (i.e., Anschauung and the Urphenomenon) were described in detail between 1794 and 1799 and explored in his *Farbenlehre*, they were alluded to in his *Beiträge*.

THE EXPERIMENTS

Goethe introduces the prism as a universally known

> oblong glass object whose two end planes are formed by equal, parallel triangles. Parallel edges emanate at right angles from the angles of both planes, connect them, and form three equal sides […] These triangles which determine the shape of the prism are usually equilateral. It follows that all angles are also equal, each being 60 degrees […] To being, let us undertake to look through a prism and observe the objects in the room and in the landscape. The angles through which one peers can be upward or downward; the prism can be horizontal or vertical. The same phenomenon will always be observed.
>
> *(§34, 35, 37; Goethe 1971, p. 21)*

Place the prism in front of your face, close to your eyes, near the bridge of your nose and look at the various objects around you. You will see coloured lines, like the spectrum of a rainbow, all around you. Goethe continues,

> I should like you observe these phenomena long enough for you yourself to sense a desire to investigate the principles behind them and to work your way of this brilliant labyrinth. Only then do I wish you to undertake the following experiments and allow yourself to follow the demonstration attentively. What was at first playful will become a serious undertaking.
>
> *(§37; Goethe 1971, p. 21)*

You "will see a diversity of colour everywhere that simultaneously repeats the rainbow in variegated ways". You "will observe especially those colours on the horizontal edges and on small objects as the liveliest ... Horizontal lines will be coloured as well as bent, yet no colours will be observed on vertical lines" (§37; Goethe 1971, pp. 22–23).

But note: these variegated colours do not show up on single coloured or on black or white surfaces, or a clear blue sky. Once another colour is introduced, these multicoloured spectrums appear. In the night sky, Goethe observes, the stars appear as coloured flames.

"To convince yourself thereof, place the reproduced card No. 1 (Figure 3.4) in front of the prism". What do you see?

FIGURE 3.4 Card No. 1 from the playing card deck, "Black worm-like grooves on a white background" ["Nr. 1. Schwarze wurmförmige Züge auf weißem Grunde"] (LA I 3, p. 450).

Goethe takes us through the phenomenon in front of us,

> … you will see how the colours cling to the worm-like drawn lines. You will observe a corresponding but indistinct and partly blurted emission of colours.
>
> *(§43; Goethe 1971, p. 25)*

Goethe takes us through all 27 cards speaking in the first person. He notes what the observer should see when looking through the prism. The *Beiträge* become personable and comforting as each phenomenon is described to us in detail as though we are having an actual conversation with the great polymath himself. Goethe shows us that his method is really an activity that can be repeated by different people hundreds of years apart.

But what of Goethe's *theory* of colour? Goethe's *Beiträge* contains no explanation as to why the colour is there or how refraction works. The work simply describes what you see when you manipulate the cards using a prism. In doing the experiments the theory [*theoria*] starts to emerge. In Card No. 3 (Figure 3.5), namely a horizontal white stripe on a black background, a rainbow appears on the upper and lower edges of the white stripe, "its colour in complete accord with one you might observe in the sky, red on top,

FIGURE 3.5 Card Nos. 4 and 3 from the playing card deck. Left: "4. A black bar on a white background" [4. Ein schwarzen Stab auf weißem Grunde] and; Right: "3. A white bar on a black background" [3. Ein weißer Stab auf schwarzen Grunde] (LA I 3, p. 450).

followed downward by yellow, green, blue, violet" (§45; Goethe 1971, p. 27). If we take Card No. 4, a black horizontal stripe on a white background, we see the colours reversed. If the cards are manipulated by turning them 90 degrees so the black stripe is vertical, we see the same phenomenon, this time at the ends of the stripe. Do the same with Card No. 3, the same phenomenon occurs but the order of colours is reversed.

Goethe follows this up with Card No. 7 (Figure 3.6), the rectangular card divided into a white and a black square. If we place the card vertically, with the black part on top and the white on the bottom, we see through the prism a red and yellow band. Reverse the card and we see blue, green or violet on the same edge.

Goethe provides a rudimentary hypothesis through his experiments. Goethe believed that we

> can find, prepare and mix coloured substances and pigments that closely represent prismatic colours. Hence pure white represents light, pure black represents darkness. White and black are not colours in the same sense that prismatic emissions are called coloured. But there are available white and black pigments good enough to allow transmission of their appearance to other substances.
>
> *(§29; Goethe 1971, p. 20)*

FIGURE 3.6 Card No. 7 from the playing card pack "7. A half black, half white panel" [7. Eine half schwarz, half weiße Tafel] (LA I 3, p. 450).

Goethe's light–dark hypothesis is no more than an observation. A completely white or black surface, for instance, will not show any colour when seen through a prism. When black and white surfaces interact, we see colour emerge through the prism, as though prismatic colour can only be experienced between darkness (black) and light (white). Through experimentation we see colour emerge when it interacts with light and darkness, thereby becoming the theory, again highlighting that "theory" in the Goethean sense is an activity.

DEVELOPMENT OF METHOD BETWEEN *BEITRÄGE* AND *FARBENLEHRE* (1793–1805)

Goethe initially intended to write four more chapters for the *Beiträge*, having abandoned the task after drafting "Concerning coloured shadows" as the third part and "Experiments to discover the element of colour" as the fourth part in 1793.[15] By 1792, Goethe seemingly became concerned with what his experiments were meant to show empirically. Did they in fact prove or disprove a theorem, or were they simply neutral? Goethe goes on,

> I consider the direct application of an experiment [Versuch] to prove a hypothesis to be harmful, and have thereby indicated that I consider an indirect application of it to be useful, and since this is the point on which everything depends, it is necessary to explain myself clearly.[16]

Remember Goethe uses the experiment to build his hypothesis (i.e., "an indirect application") rather than using it to prove (or disprove) an existing one. Goethe's hypotheses derive from practise (i.e., experimentation), namely observation. Also note that Goethe uses the term "Versuch", which can be translated as "experiment" or as an "attempt". These experiments are not contemporary with those in experimental science, where an existing hypothesis is tested through experimentation and then repeated.

> In living Nature nothing happens that is not connected with the whole, and if experiences appear to us only in isolation, if we have to regard experiments [Versuche] only as isolated facts, this does not mean that they are isolated; the question is: how do we find the connections between these phenomena, these events?[17]

If phenomena are not viewed as isolated facts, but as part of a continuum, are we able to discover the whole? If we assume that our experiments are to discover and understand the whole and the whole becomes our theory, then we can dismiss Goethe theory of colours as an experiment within a hypothesis-testing framework completely.

> In the first two parts of my *Beiträge* I have tried to set up a series of experiments [Versuche] which at first glance border on one another and directly touch one another, and when one knows them all thoroughly and summarises them, they constitute, as it were, only one experiment [Versuch], only one experience among a manifold of views.[18]

The notion of "only one experience among the most manifold views" brings up the idea of wholeness once again. "Such an experience", Goethe continues, "which consists of several others, is obviously of a higher kind". In order to perceive the whole, a certain type of experience is required, one that is beyond just looking. On 15 January 1798 Goethe wrote,

> For the observer never sees the pure phenomenon with his own eyes, rather much depends on his mental state, on the mood of the organ at the moment [...] one has to drink an ocean dry if one wants to hold on to the individuality of the phenomenon and to observe, measure, weigh and describe it.[19]

Could this pure phenomenon be "one experience among a manifold of views [...] of a higher kind" that Goethe refers to in 1792? Goethe distinguishes the pure phenomenon from,

> The *empirical phenomenon* that every human being becomes aware of in nature, and which is subsequently elevated to a *scientific phenomenon* through experiments [Versuche], by presenting it under different circumstances and conditions than those initially known, and in a more or less successful sequence.[20]

Precise and consistent definitions are not Goethe's forte. What he means by empirical phenomena are the day-to-day observations made by a trained and untrained eye. We may, for example, see the foliage of a deciduous tree change colour. The scientific phenomenon is when we dissect the leaf, place it on a slide and view it under a high-powered microscope. These observations can be repeated if the same procedures are initiated. The pure phenomenon

> ultimately exists as the result of all experiences and experiments [Versuche]. It may not always be isolated, but rather it manifests itself in a continuous sequence of phenomena. To represent it, the human mind determines the empirically fluctuating, eliminates the random, separates the impure, develops the confused, and even discovers the unknown.[21]

In 1798 Goethe defines the Urphenomenon under the term "pure phenomenon", a term he never uses again. The point of all the experiments [Versuche] is to gain an understanding of the whole. Unlike empirical and scientific phenomena, which can be photographed, drawn, dissected and measured, the Urphenomenon lives within our mind's eye, namely experiences that are recalled by a trained observer, such as a taxonomist or petrologist. If you were to conjure up a person who you know well in your mind's eye, you will see all aspects of that person; you will recall them as a child, as an adult, when they are seated or riding a bike. In no circumstances can you convey the Urphenomenon as a single image, in the same way you can't show all the developmental phases of a flower in a single image of a flower. The Urphenomenon is a way to discover the whole, and many taxonomists have used it to discover new species and genera or find new parts of an organism's anatomy. Urphenomenon is entirely absent in mathematics due to it being abstract. A mathematical equation or number, for example, cannot be experienced as a phenomenon. Goethe tried to discover wholeness in colour through the Urphenomenon by undertaking a series of experiments using a prism and playing cards in his *Beiträge*.

Wholeness provides natural objects with meaning and we achieve this via Anschauung and the Urphenomenon. Once wholeness is discovered we can talk about our phenomenon, describe its form, its behaviour and other aspects. We can categorise it within a natural classification.

Classification is not just identification, but rather it gives natural objects meaning. For instance, we can use various techniques to identify organisms (e.g., DNA barcoding), which use data that is largely meaningless to an observer (e.g., molecular data is just a seemingly endless collection of Gs, As, Cs and Ts). Just say we find an organism and we take its DNA and run it through a database. That database then informs us that the organism is a desman. While it has successfully identified the organism and given several other facts, such as that it lives in the Pyrenees, it conveys little to no meaning. We may ironically ask "What is a desman?" after it has been identified. Classification, however, provides meaning because it can tell you the desman's relationship to other organisms.

A desman is a type of mole.

Suddenly we start to see the "mole-ness" emerge from the desman. We have meaning even if we don't have a name for the organism or a database of its DNA. A mammalogist who is experienced in mole classification will immediately know if an identification is incorrect if it informs them that a desman is a type of horse. The Urphenomenon, years in the making, tells the expert that this cannot be possible. Identification is useful in many circumstances; however, classification is a way to give meaning to organisms via Anschauung and the Urphenomenon.

In the early 1790s Goethe was also working on his *Outline for a general Introduction to Comparative Anatomy, Commencing with Osteology*[22] (see Chapter 5), in which the Typus or archetype (i.e., the whole) plays an important role in biological classification. Could his *Beiträge* have sought to find a colour archetype? In order to do so, colour would need to be treated morphologically, compared and classified, which is exactly what Goethe did in his *Beiträge*.

Goethe had established a simple classification by noting that pigment is representative of light and darkness and of "coloured substances and pigments that closely represent prismatic colours". Goethe classified these as *absolute colours*, those seen through the prism or in a rainbow, and "coloured substances and pigments" as *body colours*, namely those found in paint or in print (§26; Goethe 1971, p. 20). These terms were later developed in 1798, in an exchange with Schiller, into three colour categories, namely,

1. Physiological, namely those in relationship to the eye
2. Physical, namely those in relationship to light and darkness[23]
3. Chemical, namely those in relationship to matter on which they appear.[24]

We will investigate this colour classification later. I only include it to show that Goethe's colour experiments led to the discovery of natural colour classification in which any phenomenon falls into one of three categories. The same is true of biological classification. A vertebrate with lactating glands and hair is always more closely related to other mammals than it is to any other organism. In this sense, natural classification (i.e., the whole) is greater than the sum of its parts. Goethe's study in colour is no different than studying comparative anatomy, something which he was doing concurrently as he was pondering over during the transition period between the *Beiträge* and *Farbenlehre*.

Goethe outlined his method in 1805, which "expresses a central theme in Goethe's scientific writing".[25] The short essay expresses the importance of Anschauung and Urphenomenon in his method.

> Two demands arise in us when we observe natural phenomena: to know the phenomena themselves completely and to reveal them through reflection. Order leads to completeness; order demands method; and method unburdens [erleichtert] the mental picture [Vorstellung]. When we survey an object in all its parts, grasp it correctly, and can recall it in our mind's eye, then we may say that we perceive [anschauen] it in the true and higher sense, that it belongs to us, that we attain a certain mastery over it. And thus, the particular always leads us to the general, and the general to the particular. Both influence each other in every observation, in every presentation.[26]

Goethe is clearly discussing the Urphenomenon but provides an extra detail: "the particular always leads us to the general, the general to the particular. Both interact with each other in every observation, in every statement". Within an Urphenomenon, say the lifetime of a pet bird you have reared from an egg, you will remember when it was an egg and snapshots of it becoming an adult. The appearance of an egg becoming a fully grown chicken is what Goethe termed Steigerung, namely a rise, growth or development. At either end of Steigerung lies the general (an egg) and the particular (the mature bird with all its characteristics and behaviours).

The Urphenomenon leads us to a natural classification and within natural classifications we have elements that lead us from general types to specialised types (often confused in biology as "primitive" and "derived" in context to archetypes). Yet, archetypes are rarely shown or described in taxonomy; in fact no taxonomic paper or monographic treatment has a Methods section that describes the Urphenomenon that lies within the author's mind's eye. The reason is the impossibility of relating the Urphenomenon in words or symbols, which are "based on intuitive perceptions alone and cannot occur in language".[27]

By 1807 Goethe is publishing on Anschauung and wholeness in his zoological and botanical work, indicating that his morphological method and colour theory overlap to the extent that they are the same thing. Compare the above passage with the following.

> Thus scientific minds of every epoch have also exhibited an urge to understand living formations as such, to grasp their outward, visible, tangible parts in context, to see these parts as an indication of what lies with and thereby gain some understanding of the whole through an exercise in Anschauung.[28]

By 1807 Goethe had established "his" method, which he admits is an existing method used in natural history. The terms Anschauung and wholeness were already in use by the polymath; only Urphenomenon was yet to be coined.

GOETHE'S ISSUE WITH NEWTON

Boyle states "the adoption of a theory hostile to Newton was of course part of [Goethe's] original insight".[29] Particularly when Goethe noted, "The power to produce colours is independent of refraction".[30]

Goethe took issue with the Newtonian notion that refracted white light is the source of colour.[31] In his own observations using a prism (as discussed above) Goethe saw that white surfaces only produced colour when it interacted with another turbid medium, such as a dark border. A white wall remains white, so where are the colours that refraction promises?

> There is a dilemma here, almost similar to whether the Sun rotates around the Earth or whether the Earth rotates around the Sun, given both can more or less explain the phenomena [of sunrise and sunset]. If one accepts the Newtonian approach [refraction as the basis of colour] and begins with the coloured image, then the white space is very difficult to explain. If one begins with the coloured edges, then one takes the white space for what it is, for unaltered light, and the green from yellow and blue, the purple from blue and yellow-red, is explained very nicely by mixing it together.[32]

Many critics of Goethe's *Beiträge* and *Farbenlehre* have spilt much ink in detailing his argument in painstaking detail.[33] His surprise at not seeing a myriad of colours emanating from a white surface led Goethe to start writing his *Beiträge* possibly in the winter of 1790/1791 in Gotha.[34]

Rudolf Magnus noted that Goethe's *Beiträge* "took one important step forward. He divided all colour phenomena into three great groups [i.e., physiological, physical and chemical colours]" and that his "findings celebrated their greatest triumph in explaining a phenomenon [i.e., coloured after-images] that had long been known but always misinterpreted" (Magnus 1949, pp. 134, 155). Why, then, is there this discrepancy between Goethe's critics?

The problem is with Goethe's Polemic Part in his *Farbenlehre*. The Polemic was a brazen attempt to use his own method to disprove Newton's theory, which to a small extent made a salient point, yet the victory was Pyrrhic. Goethe had made several political errors of his own, mostly in his Polemic, thereby condemning his entire *Farbenlehre* to the historical dustbin. Yet *Farbenlehre* included an excellent Didactic Part and possibly the first history of science in his Historical Part. Before we explore *Farbenlehre* it is important to state why Newton irritated Goethe.

Newton's work attempted to explain the mathematics behind light rays and in doing so used colour as an example. What Newton did not set out to do is classify colour morphologically and provide meaning.

Goethe saw his approach as different to Newton's, which at the time was widely adopted and ignored Anschauung and the Urphenomenon. Newton was not interested in describing colour morphologically, nor was he interested in finding a natural classification (i.e., a whole), into which all colour phenomena exist (i.e., physiological, physical and chemical).

The *Beiträge* describes how we *experience* the emergence and behaviour of colour in light and darkness. If we were to observe an animal or plant, we would observe morphology over time and take note of the organism's form, ontology and behaviour.

Here we can classify the organism and compare it to other organisms and so forth. The observation and the classification not only add to our knowledge of an organism but also bring meaning. Grand explanations, such as the orbit of the Earth around a stationary Sun, offer little meaning to someone watching the Sun move as it rises and sets. If we, for example, were to show Goethe a stuffed platypus and say, "it's from New South Wales", little meaning about the organism is conveyed. The same is true if we said that the colour red has a dominant wavelength of 625 to 750 nanometres.[35] How does wavelength convey meaning to an observer? How can an artist derive meaning from wavelength when using red pigment in a painting? Goethe acknowledged that Newton's theory was about providing a mathematical explanation as to how rays form the colour red (i.e., wavelength of 625 to 750 nanometres). Newton's theory does little to explain how we experience the colour red through a prism or in a painting and its relationship to lightness and darkness. Goethe saw a gap in this knowledge and sought to investigate it further through his physiological colours, providing a method that Magnus called "an epoch-making work" (Magnus 1949, p. 199).

I will not explore Goethe's critique of Newton (i.e., the Polemic Part) any further as this has been done in greater detail by Sepper (1988), with an English translation provided by Michael Duck and Michael Petry (in Goethe 2016). The Polemic Part provides little insight into Goethe's method;[36] in fact, it distracts us from appreciating what he had achieved as a scientist. Why Goethe decided to write the Polemic Part and doggedly pursue Newton has been considered as psychological, theological or phenomenological.[37] Goethe did himself no favours with his anti-Newtonian stance; in fact, it has overshadowed his entire method and his botanical and zoological work. In "The Translators Preface", Charles Lock Eastlake (1793–1865) sums up Goethe's predicament.

> Had he contented himself with merely detailing his experiments and showing their application to the laws of chromatic harmony, leaving it to others to reconcile them as they could with the pre-established system, or even to doubt in consequence, the truth of some of the Newtonian conclusions, he would have enjoyed the credit he deserved for the accuracy and the utility of his investigations. As it was, the uncompromising expression of his convictions only exposed him to the resentment or silent neglect of a great portion of the scientific world, so that for a time he could not even obtain a fair hearing for the less objectionable or rather highly valuable communications contained in his book.
>
> *(Eastlake in Goethe 1840, p. xxix)*

Goethe's Art of Science aims to outline Goethe's method and not what he thought or believed. Scientists often have their own pet theories, yet it has little influence on how they do their science. An 18th-century creationist taxonomist and a 21st-century comparative biologist would still describe and diagnose a new species in the same way regardless of their beliefs. Should we reject Linnaeus' taxonomy system because he didn't believe in evolution?

Goethe's method was about how we observe phenomena and not about abstract wave or particle theory. Regardless of what we believe about the immutability of light, we still see blue emerge out of black (i.e., darkness) and yellow emerge from white (i.e., lightness) when we look through a prism. Goethe was concerned about how we classify those phenomena, namely the ones we experience as observers. I am not defending Goethe's belief in the immutability of light but rather I champion his method, one that I and my taxonomist colleagues have used time and time again in our own observation of the natural world.

ZUR FARBENLEHRE (1808–1810)

Goethe's *Zur Farbenlehre* (1808–1810) is divided into three parts:

> 1. The Didactic Part[38] "offers a theory in outline. There are many examples of colour phenomena that are categorised by general type, and the categories are presented in an order explained in the Introduction [of the Didactic Part]"
> 2. The Polemic Part[39] "will undertake to lift the veil from Newton's theory, which by virtue of its power and authority has long obstructed an unprejudiced view of colour phenomenon"
> 3. The Historical Part[40] "will be devoted to historical investigations and the work of earlier commentators. Previously we expressed the thought that the history of a person portrays that person; it is also possible to say that the history of science is science itself".
>
> *(Goethe 1995, pp. 159–160)*

The parts are not equal in length. The Polemic Part (LA I 5, pp. 1–195) constitutes the smallest section, the Didactic Part (LA I 4, pp. 1–266) is the next largest and the Historical Part (LA I 6, pp. 1–429) is the largest. In this section I will discuss Goethe's colour classification further, skip the Polemic Part for the reasons given above and continue with the Historical Part, the latter of which remains mostly untranslated.[41]

The underlying maxim of Goethe's method is briefly outlined in the Introduction to the Didactic Part,

> ... we find that many would prefer to dismiss phenomena with a general precept or a quick explanation with taking the trouble to study them in detail and achieve a knowledge of the whole over a longer period of time.[42]

In other words, don't put the cart (i.e., scientific model) before the proverbial horse (i.e., the phenomenon). In doing so the phenomenon is often overlooked in favour of the model. If, for instance, we only see the distribution of the Permian fossil plant *Glossopteris* as the result of former land-bridges between continents or long-distance dispersal, Alfred Wegener would never have come across the idea of Continental Drift and the later discovery of Plate Tectonics (Frankel 2012, p. xix). Entrenching the idea that Earth's continents were fixed, regardless of any conflicting palaeobiogeographical evidence, highlights the problem of an overarching model that ignores evidence. The problem still occurs today, both in science and in society. US mathematician Cathy O'Neil states that models, regardless of how well-informed they may be, "are opinions embedded in mathematics" (O'Neil 2016, p. 21). Goethe had not seen this in his own geological investigations, where he did precisely that: treat the Wernerian Plutonist model (i.e., well-intended opinions) as informing us about the evidence, rather than the other way around. Viewing natural phenomena such as rocks and minerals through Wernerian rose-coloured glasses led Goethe to making very rudimentary errors in rock and mineral classification. Goethe's insistence to "let the phenomena speak for themselves" had been overlooked by the polymath in geology, but did he do the same in his *Farbenlehre*?

Colour, unlike rocks and minerals, can be manipulated so it emerges every time we look through a prism. Other products of Nature are historical and their development may not be observed emerging as it were from a magmatic melt. Rocks, minerals and fossils are historical objects whose natural generation is generally unobserved. We can't see a schist or a granite form, as the conditions are fatal to human life. We can't see joints forming in a granitic tor because we live short lives. Historical science is limited and we can only see its products rather than the processes that created rocks and minerals. In this sense it is not an experimental science. Neither is Goethe's *Farbenlehre*. Unlike the formation of granitic tors, we can see the life cycle of a chicken or potato plant. We can see the plant emerge from a seed or a chick emerge from an egg. The same is true with colour. Through a prism we can see violet emerge from the black part of playing card. Seeing colour emerge from under a prism is not a process in the sense of quantum mechanics or genetics – complex mechanisms that are only understandable using abstract measurements that do not involve human observation. While these processes may explain how photons or cells work, they do not help us to understand the phenomenon we see before us. As we look at light under a prism, colour emerges before our very eyes. To attain meaning from this phenomenon we attempt to describe what we see in context to other colours and related phenomena. To do so we need a colour classification and Goethe proposed three categories (see Table 3.1): 1. Physiological colours, 2. Physical colours and 3. Chemical colours. "The first are entirely ephemeral, the second are transient but always linger for a time, and the third may be held constant over long periods of time".[43]

TABLE 3.1
An Attempt to Simplify Goethe's Classification of Colour of 1810

1. Physiological Colours	2. Physical Colours	3. Chemical Colours
I. Light and dark as sensed by the eye	**XI. Dioptric colours**	**Colour Circle**
II. Black and white forms as sensed by the eye	X. Dioptric colours of the first class	
III. Grey surfaces and forms	Entoptic colours (1820)	
IV. The blindingly bright colourless form	XI. Dioptric colours of the second class (refraction)	
V. Coloured forms	XXXI. Catoptric colours	
VI. Coloured shadows	XXXII. Paroptic colours	
VII. Lights producing a weak effect	XXXIII. Epoptic colours	
VIII. Subjective halos		
Supplement: Pathological colours		

Note: original chapter and section numbers have been retained.

Physiological colours, Goethe notes, "are wholly, or largely, the property of the observer",[44] something that was, until then, considered to be merely illusions and inconsequential. What, then, do we see? Goethe insists that we see light and dark. In the latter case,

> with our eyes open in a completely dark room we sense a certain deprivation. Our organ of sight is left to itself, it withdraws into itself and loses the stimulation, fulfilling contact which unites it with the outer world and makes it whole.

Doing the opposite, exposing it to bright white light, our eyes "are blinded and for a while unable to distinguish less brightly illuminated objects".[45] We experience these little "experiments" almost every day, yet we make little of them, as we take our ability to see and perceive as granted. Yet Goethe is doing something new. He is trying to show that it is *you* and *your eyes* that are the scientific instruments recording and measuring the phenomena.

Goethe provides further situations in which we (our eyes) sense light and darkness through the process of seeing [sehen]. "At one and the same time we perceive all the immediate degrees of light and shadow [Helldunkel], and all the distinct qualities of colour [Farbenbestimmungen]".[46] We may jump to the conclusion that Goethe is talking about the Urphenomenon, namely our ability to recall the phenomenon via our mind's eye and past experiences. Goethe is in fact describing what we actually perceive within a single event, such as the René Magritte painting *The Empire of Light* (1954), in which we have an image of a house behind some trees at night, with coloured shadows lit by a solitary lamp that gives off dim glow. At the same time the sky is lit by daylight, giving it a bright light blue colour with white fluffy clouds. You can perceive light, darkness and a wide range of colour in the one image. While all these are confined to a painting, you can easily experience this during a thunderstorm when dark clouds cast darkness across a landscape, but you can still see the blue sky in the distance (Figure 3.7). At this point in Goethe's classification of physiological colour he intends

> to consider and note these elements of the visible world [the immediate degrees of light and shadow, and all the distinct qualities of colour], and how our organ of sight responds to them. With this as a goal we will begin by considering the simplest type of form [Bilder].[47]

The term "form" sounds resoundingly like Morphology and seems out of place on a treatise on colour. What Goethe may be referring to here is "shape" or "image" rather than as a natural object (i.e., organisms and minerals); for example, "[w]e look from one object to the next, and the succession of [shapes/images] seems clear-cut".[48]

The Didactic Part of *Farbenlehre* is similar to that of the *Beiträge*, in that Goethe takes us on a journey through various situations in which we may have experienced the same phenomena that he describes in detail. One difference is the historical context Goethe provides, citing previous authors, such as Tycho de Brahe (1546–1601) and Giovanni Battista Beccaria (1716–1781), who have noted the same phenomena. Later, Goethe returns to his "experiments" or "attempts" [Versuche].

FIGURE 3.7 An example of Epoptic colours on a transparent solid, namely a soap bubble. (Photo by Julie Martin used with permission.)

> Let us hold a small piece of brightly coloured paper or silk against a moderately illuminated white background, look steadily at the small coloured surface for a time, and then remove it without shifting our gaze. A spectral image in a different colour will appear against the white background. We may also leave the coloured paper in place and look at a different spot on the white background to observe the appearance of this colour, for it arise from [an image] which now belongs to the eye.[49]

The phenomena encountered in the above experiment neatly segues into Goethe's Colour Wheel or more correctly Colour Circle [Farbenkreis].

> As a quick aid to finding the colours produced by this activity of opposition we may refer to the illustration of the colour [circle] in our [plate (Figure 3.8)]. This [circle] conforms fully to nature in its arrangement and will help in the present discussion because the colours placed diametrically opposite one another on the [circle] are those which demand each other in the eye as complements. This yellow demands violet; orange, blue; purple, green; and vice versa. Thus each graduation requires its reciprocal, the simpler colour requires the more complex, and vice versa.[50]

Goethe's colour wheel is based on experience of colour, namely, the after-image of a colour. It would be folly to compare Goethe's and Newton's colour circles. Newton's

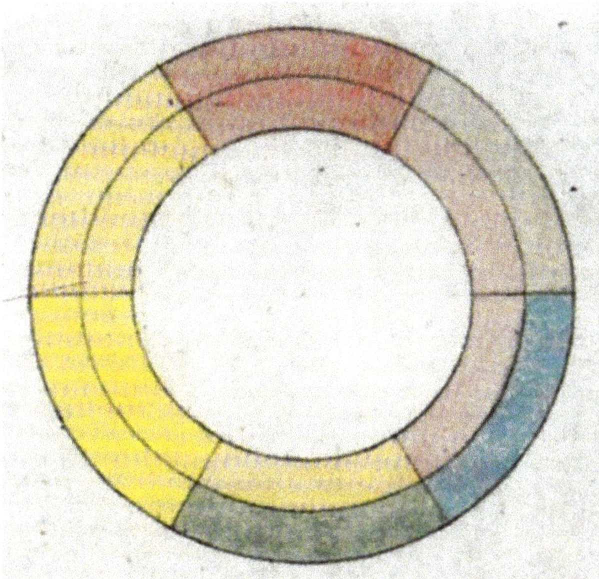

FIGURE 3.8 Goethe's colour circle or Farbenkreis (LA I 7, p. 45, Plate I, 1).

colour circle is based on a geometric model in which light rays of a certain colour cancel each other out, forming white light.[51] Goethe's colour was not a model; rather, it is based on our experience of colour and the effect of coloured after-images. While Newton's wheel was grounded in geometry, with the need for two coloured rays to cancel each other out, Goethe's method required our eyes and our mind. "The eye", Goethe says, "exhibits an exacting need for wholeness" and "completes the circle of colours within itself".[52]

Goethe's Physiological colours cover further phenomena including "Coloured shadows", "Lights that produce a weak effect", "subjective halos" and "Pathological colours", namely colours that people with certain conditions see unaided (i.e., colour blindness).

The second section of the Didactic Part, "Physical colours", is divided into dioptric colours of the first and second class.

> The term dioptric is applied to those colours which require a colourless medium for their creation, with the stipulation that light and dark must pass through the medium and affect either the eye or a surface standing opposite. It is therefore necessary that the medium be transparent or at least translucent to some degree.[53]

Phenomena in the first class are produced by turbid media while the second are those phenomena produced by highly transparent media. The idea is that colour, as seen through a mildly turbid media, is yellow and as turbidity increases the colour turns yellow-red then ruby-red. Goethe notes the same colours occur when the sun is seen through a haze, creating sublime sunsets and sunrises. The same is true when looking at darkness. Through the turbid media the colour appears blue and with increased turbidity it becomes light blue. Goethe reminds us that this also occurs when looking out into mountainous landscapes, which appear blue, but with an increase in haze, the light appears light blue within the valleys. The same phenomenon can be seen in candle flames. When a candle flame is held against a dark background you can see the appearance of blue, which disappears when you hold the flame against a white background. Goethe coined these *Urphenomena*.[54]

It is in the Didactic Part where Goethe finally outlines his method.

> In general, events we come aware of through experience are simply those we can categorise empirically after some observation. These empirical categories may be further subsumed under scientific categories leading to even higher levels.

It is assumed that the observer is trained in some way to describe and identify an organism. That organism is observed and categorised empirically; that is, we can place it into a taxonomic group, say macropods (kangaroos). What we are observing is some kind of macropod. We then compare it to the macropods we have previously seen and ones we may have in our national natural museum collections or zoos.

> In this process we become familiar with certain requisite conditions for what is manifesting itself. From this point everything gradually falls into place under higher principles and laws revealed not to our reason through words or hypotheses, but to our [Anschauung] through phenomena.

What is manifesting itself is something we are familiar with even though this might be first time we are seeing the phenomenon. Many tourists I meet visiting the Blue Mountains in New South Wales see wallabies and call them kangaroos, in the sense of *Macropus*, the red and grey kangaroos made famous by Skippy or the boxing kangaroo. While it's a good guess, wallabies are much darker in colour, stockier and shorter than a red or grey kangaroo. Even so, the kangaroo-ness of the organism had manifested itself and with the help of previous encounters, perhaps at the zoo. In this sense the kangaroo emerges through the Urphenomenon,

> We call these Urphenomena because nothing higher manifests itself in the world; such phenomena, on the other hand, make it possible for us to descend, just as we ascended, by going step by step from the [Urphenomenon] to the most mundane occurrences in our daily experience.[55]

Nothing higher can manifest itself and anything that does is pure fiction or speculation based on no empiricism whatsoever. In other words, it's made up. Yet the Urphenomenon of kangaroos is unlimited in that we can always observe more of that organism, including its behaviour and anatomy. For a mammal taxonomist, who is trained in the anatomy of kangaroos, these observations are highly specialised. Even so, we still function as normal people in observing the world in our day-to-day lives. Yet, Goethe notes, we place secondary phenomena in a higher place with science. Secondary phenomena are those colours seen through clear media (think of the glass prism), which don't really occur in nature, with the exception of some minerals (see below).

> But even when we find such an Urphenomenon, a further problem arises when we refuse to recognise it as such, when we seek something more behind it and above it despite the fact that this is where we ought to acknowledge the limit of our perception.

Essentially what Goethe is saying is that inferences, speculation and second-guessing are not part of our knowledge, namely, the Urphenomenon. We may acknowledge that we have through Anschauung discovered a new type of fossil mammal. We can't say that that organism is the ancestor of all kangaroos, because we simply do not possess any Anschauung to confirm it is true.

> It is proper for the natural scientist to leave the [Urphenomenon] undisturbed in its eternal repose and grander, and for the philosopher to accept it into [their] realm. There [they] will discover that a material worthy of further thought and work has been given him, not in individual cases, general categories, opinions and hypotheses, but in the basic phenomenon and Urphenomenon.[56]

The Urphenomenon is much more empirical in nature than any species (i.e., category), identified ancestor (i.e., opinion) and Newtonian colour circle (i.e., hypotheses). The only problem is that it is purely descriptive in nature, and while it does convey meaning (e.g., a desman is a type of mole; yellow emerges from light), it lacks the speculation and story-telling that science and the science-loving public crave. Surely "The ancestor to all kangaroos has been found" sounds far more thrilling and captivating than "Wallabies are a type of kangaroo"?

The Didactic Part describes the categories of Catoptric, Paroptric and Epoptic colours. Catoptric colours are "those colours which appear as the result of reflection",[57] whereas Paroptric colours occur "when light passes across the edge of a colourless opaque object"[58] and are related to Catoptric colours because they both refract light, making them secondary colours (i.e., Dioptric colours of the second class). Epoptic colours "arise on the surface of a colourless object, directly, without out mediation, dye, or immersion".[59] These colours can occur on solid objects when two pieces of glass or lenses are brought together, or on broken surfaces, or along cracks (i.e., cleavages) within transparent minerals. In the case of liquids, it's the colours created on soap bubbles, film on liquids or from our breath on a glass surface. While these colours are ephemeral they do form a transition from Physical to Chemical colours.

Chemical colours, that is, the colours within the Colour circle, are those colours that are properties of objects. These colours can fade, wash away and so on, but they are mostly permanent (i.e., paint or the colour of yellow quartz). Goethe provides few experiments and mostly examples (some hearsay) in this section of the Didactic Part. Goethe provides examples from the animal, plant and mineral kingdoms. The fourth section of the Didactic Part is on "General observations on the principles of colour",[60] which returns the reader to the Farbenkreis and its application to colour mixing – "opposite phenomena do not cancel themselves out, but join in a third phenomenon",[61] namely, a new colour (i.e., blue and yellow give you green). The fifth section discusses colour in relationship to other fields, such as mathematics.

> The theory of colour, in particular, has been hurt and greatly hindered in its progress by being lumped with the area of optic dependent of geometry. It may, in fact, be considered entirely separate from geometry.[62]

And its relationship to general physics is discussed.

> The present state of general physics seems especially favourable for our work; constant and wide-ranging research have brought [natural history] to such a high level that it now seems possible to relate the endless realm of empirical phenomena to one central method [i.e., Goethe's method].[63]

Goethe's method, namely Anschauung and the Urphenomenon, was the main goal in his *Farbenlehre*.

> We, too, have long wished to introduce the term *polarity* into the theory of colour, and the present work will show our justification and purpose in doing so. Later on we may have an opportunity to link the elementary phenomena of nature in our own way by using this approach, this symbolism always accompanied by the [Anschauung] belonging to it.[64]

As I mentioned above, in 1805 Goethe penned a small essay titled *Polarity*[65] in which he again describes the "pure phenomenon", later to be coined the Urphenomenon. In the piece Goethe attempts to show that the Urphenomenon has polar extremes (i.e., the egg on one end and the adult chicken on the other), that is, the limit of our perception and the phenomenon.[66] In Goethe's method polarity is the hypothesis of wholeness and the Urphenomenon is the outcome. Goethe proposed that colour emerging from darkness and light within a turbid medium was the hypothesis, the Anschauung, that we employ during the experiments that we conduct with our prism, and the ability to recall the polarity in our mind's eye was the result. As such, polarity would make sense to explain the phenomenon of colour,[67] as Goethe does in this Didactic Part.

The final section of the Didactic Part concerns itself with the "Sensory-moral effect of colour".[68] This section is the most holistic part of the *Farbenlehre*, where Goethe subscribes colour to emotions, cultural practises through history and art. The section is divided into the colours of the Farbenkreis, namely, yellow, red-yellow, yellow-red, blue, red-blue, blue-red, red (purple) and green. Goethe also explores the effects of colour combinations, colouration of objects, tone and various other notes. Goethe admitted to "publishing it in a rather sketchy form – on the spur of the moment, so to speak",[69] which shows by the end of the Didactic Part. Importantly, Goethe's method is fully developed in the Didactic Part of *Farbenlehre* and not further investigated in his Polemic and Historical Parts.[70]

FURTHER WRITINGS OF COLOUR (1813–1832)

Goethe expanded his classification of colour through the discovery of Entoptic colours, "which appear fleeting or permanent on the surface of bodies under various conditions".[71] Entoptic colours are phenomena that embrace polarity.

> When we attribute polarity to entoptic phenomena, we do so in the sense in which we have endeavoured to develop all chromagenesis [colour production] in our theory of colour. Darkness and light are primordial opposites, eternally alien to one another; only the matter that lies within and between the two, when it is physically opaque, has an illuminated and a dark side; however, only in weak backlight does the shadow appear.

> If the matter is translucent, what we call colour develops within it in the chiaroscuro [Helldunkel] and opaque, as seen by the eye.
>
> This, as well as light and dark, manifests itself in polar opposites. They can be suspended, neutralised, and featureless, so that both appear to disappear; but they can also be reversed, and this reversal is generally the most delicate thing in the world, with any polarity. By the slightest condition, the plus can be transformed into a minus, the minus into a plus. The same applies to entoptic phenomena. At the slightest stimulus, the white cross is transformed into a black cross, the black into a white cross, and the accompanying colours are likewise reversed into their required opposites. But it is our duty at present to break this down. Do not let go of the central concept, and despite all the variability, the basic phenomenon will always be found.[72]

At first one might confuse Entopic with Epoptic colours as both are derived from the surfaces of objects. The difference, while subtle, does explain polarity. Epoptic colours are the actual colours found on a colourless surface, such as a blue liquid dye running down a white surface in a clear medium. Imagine a tin of blue paint dashed onto a white wall in a well-lit and shadowless room. Entoptic colours occur to our eye when there is a shift between light and darkness through a turbid medium (i.e., shadows; Figure 3.9). In the first case there is no shift in colour perception. The wall

FIGURE 3.9 Entoptic colours are shown in an image of a cross. Note the darker colour along the edges of the white cross.

(Photo by the author, 2026)

remains white and the paint blue. There is no turbidity, that is, no particles that can cause shadows that shift the colour's hue. Epoptic colours are merely physical colours of the second class, and Entoptic colours are physical colours of the first class.

Polarity comes into play is when we shift the light to create coloured shadows. The white cross shifts from pure white to an array of hues (e.g., blues, greys) until it is completely black. There we have the polarity, namely,

|+|white|blue|blue – grey|grey|dark grey|black|–|

We can witness the polarity (i.e., white and black) and all the colours in-between. In this sense polarity is the Urphenomenon, which embeds itself into our mind's eye where we can recall the phenomena again and again via past experiences. In the example above, we witness the Urphenomenon in one sitting. In other cases, we can witness it over longer periods of time where we only see snapshots of the whole, for example,

|+|egg|chick|pullet|laying hen|mature hen|–|

While we are unable to see this in one go, we can see it in our mind's eye via the Urphenomenon.[73] We can visualise the egg turning into a chick, the chick into a pullet and so on. We can also visualise the mature hen turning into a laying hen, and the laying hen into a chick and so on. The more we observe the phenomenon, the more of its gradation or transformation we see, thereby attaining a wholeness of sorts. While we can watch the complete transformation of a white object becoming dark, we are not able to see the complete life cycle of a hen. In this way the hen's life as a complete whole will always be bigger than the snapshots that we see from time to time. If the Urphenomenon represents the whole it can never attain anything more. As Goethe warned Eckermann,

> …[the Urphenomenon] cannot grant him anything higher, and he should not seek anything further behind it; here is the limit. But for people, the sight of a Urphenomenon is usually not enough; they think there must be more, and they are like children who, after looking into a mirror, immediately turn it over to see what is on the other side.[74]
>
> *(Goethe to Eckermann 18 February 1829)*

Goethe's method, namely Anschauung, the Urphenomenon and the idea of wholeness, fell on deaf ears. *Farbenlehre* had many detractors, such was Hermann von Helmholtz (1821–1894)[75] who dismissed it as non-experimental. The science of colour was seen to fall into the realm of physics, which was considered experimental. Yet not all science is experimental; in fact, much of it is descriptive or historical (see Wilkins and Ebach 2014). In historical science, Goethe's method does make much sense, but these sciences (i.e., comparative biology, petrology) were in their infancy during his lifetime. Goethe was the first to openly describe the method of observation in historical science and the first to show how it is used. As Rudolf Virchow was to eulogise in 1858, "Should we (scientists) not be ashamed to confess that it was Goethe who had been obliged to rescue, for natural science, the true principles of observation?" (Virchow 2008, p. 29).

NOTES

1 Auf Alles was ich als Poet geleistet habe [...] bilde ich mir gar nichts ein.

> Es haben treffliche Dichter mit mir gelebt, es lebten noch Trefflichere vor mir, und es werden ihrer nach mir sein. Daß ich aber in meinem Jahrhundert in der schwierigen Wissenschaft der Farbenlehre der einzige bin, der das Rechte weiß, darauf tue ich mir etwas zugute, und ich habe daher ein Bewußtsein der Superiorität über Viele.
>
> *(Goethe to Eckermann, 19 February 1829,* Gespräche *II, p. 307; translation in Sullivan 2017, p. 115)*

2 "Historians of the nineteenth-century science, the really serious historians, usually dismiss anything sounding like Romantic science as an aberration and suspect anyone following such a red thread will be travelling down a path that terminates in the higher nonsense" (Richards 2002, p. 3). Richards cites Edward Wilson and Timothy Lenoir as two modern researchers who share such a view. Regardless, Richards himself labels Goethe's Morphology "arose out of a decidedly Romantic sensibility" (Richards 2002, p. 2).

3 As Sepper points out

> ... scientists were more concerned to save Newton's theory than the phenomena [and] the authority on Newton was generally considered weightier than the testimony of faithful and attentive observers [...] Scientists had been blinded by the authority of Newton, a sociological factor, and by their own confusion of theory and fact.
>
> *(Sepper 1988, pp. 102–103)*

4 Es gibt eine zarte Empirie, die sich mit dem Gegenstand innigst identisch macht und dadurch zur eigentlichen Theorie wird. Diese Steigerung des geistigen Vermögens aber gehört einer hochgebildeten Zeit an.

> Das Höchste wäre, zu begreifen, daß alles Faktische schon Theorie ist. Die Bläue des Himmels offenbart uns das Grundgesetz der Chromatik. Man suche nur nichts hinter den Phänomenen; sie selbst sind die Lehre.
>
> Wenn ich mich beim Urphänomen zuletzt beruhige, so ist es doch auch nur Resignation; aber es bleibt ein großer Unterschied, ob ich mich an den Grenzen der Menschheit resigniere oder innerhalb einer hypothetischen Beschränktheit meines bornierten Individuums.
>
> *(LA I 11, pp. 356–358)*

5 Oxford English Dictionary Online https://www.oed.com/search/dictionary/?scope=Entries&q=theory, accessed 13 December 2024.

6 Duck and Petry (in Goethe 2016, p. 8) also translate *Lehre* as Doctrine.

7 "zusammenfassend, der Gesamtinhalt eines Unterrichts, das was in einem Wissenszweige zusammenhängend gelehrt wird. Diese Bedeutung, die häufig nahe an die erste anrührt oder auch in ihr verläuft (vergl. die Fälle unten d), erscheint in mehrfachen Verbindungen" (Deutsches Wörterbuch von Jacob Grimm und Wilhelm Grimm: https://woerterbuchnetz.de/?sigle=DWB&lemid=L03578).

8 *Lehre* is also "Apprenticeship" as in "to serve an apprenticeship" ["eine Lehre machen"]; and as "Doctrine" as in "the Christian doctrine" ["die christliche lehre"], or as "Advice" or "Warning" as in "let that be a lesson to you" ["lass dir das eine Lehre sein"].

9 "... wenn die Newtonische Lehre leicht zu lernen war, so zeigten sich bei ihrer Anwendung unüberwindliche Schwierigkeiten. unsere lehre ist vielleicht schwerer zu

fassen, aber alsdann ist auch alles getan; diese Farben, welche das Fundament der ganzen Lehre machen" (https://woerterbuchnetz.de/?sigle=DWB&lemid=L03578).

10 Goethe does employ the modern usage of theory when referring to Newton's *Opticks*, "… because the theory is abstract and its application cannot be judged without the most exact insight into arithmetics" ["… weil die Theorie abstrakt ist und die Anwendung derselben ohne die genauste Einsicht in die Rechenkunst nicht beurteilt werden kann" (§13; WA II 5.1, p. 9; LA I 3 p. 10). Here we can see the two distinct usages of the term "theory" in Goethe's work.

11 https://en.langenscheidt.com/german-en.lish/lehre

12 Sepper also finds the rendering "theory of colours" problematic and prefers to refer to the "multivolume work *Zur Farbenlehre*" (Sepper 1988, p. 201, fn 1).

13 Matthaei notes "Goethe, for good reason, has been called the founder of a new investigation into phenomena" (Matthaei in Goethe 1971, p. 6). I disagree. As a practising scientist and someone who has done taxonomy, I would say that Goethe was the first scientist to make us aware of the method that we as empirical observers use.

14 Meine Absicht ist: alle Erfahrungen in diesem Fache zu sammeln, alle Versuche selbst anzustellen und sie durch ihre größte Mannigfaltigkeit durchzuführen, wodurch sie denn auch leicht nachzumachen und nicht aus dem Gesichtskreise so vieler Menschen hinausgerückt seien. Sodann die Sätze, in welchen sich die Erfahrungen von der höheren Gattung aussprechen lassen, aufzustellen und abzuwarten, in wie fern sich auch diese unter ein höheres Prinzip rangieren. (LA I 3, p. 295; translated by Douglas Miller 1995, p. 21)

15 "Der Beiträge zur Optik Drittes Stück: Von die Farbigen Schatten" LA I 3, pp. 64–81; "Versuch die Elemente der Farbenlehre zu entdecken. Der Beiträge zur Optik Viertes Stück" LA I 3, pp. 190–209.

16 Ich habe vorhin gesagt, daß ich die unmittelbare Anwendung eines Versuchs zum Beweis irgendeiner Hypothese für schädlich ansehe, und habe dadurch zu erkennen gegeben, daß ich eine mittelbare Anwendung derselbe für nützlich halte, und da auf diesen Punkt alles ankommt, so ist es nötig, sich deutlich zu erklären. (LA I 3, p. 292)

17 In der lebendigen Natur geschieht nichts, was nicht in einer Verbindung mit dem Ganzen stehe, und wenn uns die Erfahrungen nur isoliert erscheinen, wenn wir die Versuche nur als isolierte Fakta anzusehen haben, so wird dadurch nicht gesagt, daß sie isoliert seien, es ist nur die Frage: wie finden wir die Verbindungen dieser Phänomene, dieser Begebenheiten? (LA I 3, p. 292)

18 Ich habe in den zwei ersten Stücken meiner optischen Beiträge eine solche Reihe von Versuchen aufzustellen gesucht, die zunächst aneinander grenzen und sich unmittelbar berühren, ja wenn man sie alle genau kennt und übersieht, gleichsam nur einen Versuch ausmachen, nur eine Erfahrung unter den mannigfaltigsten Ansichten darstellen. (LA I 3, p. 293)

19 Denn da der Beobachter nie das reine Phänomen mit Augen sieht, sondern vieles von seiner Geistesstimmung, von der Stimmung des Organs in Augenblick, von Licht, Luft, Witterung, Körpern, Behandlung und tausendandern Umständen abhängt; so ist ein Meer auszutrinken, wenn man sich an Individualität des Phänomens halten und diese beobachten, messen, wägen und beschreiben will. (LA I 3, p. 307)

20

1. Das empirische Phänomen, das jeder Mensch in der Natur gewahr wird, und das nachher
2. zum wissenschaftlichen Phänomen durch Versuche erhoben wird, indem man es unter andern Umständen und Bedingungen als es zuerst bekannt gewesen, und in einer mehr oder weniger glücklichen Folge darstellt.

(LA I 3, pp. 307–308)

21

> 3. Das reine Phänomen steht nun zuletzt als Resultat aller Erfahrungen und Versuche. Es kann niemals isoliert sein, sondern es zeigt sich in einer stetigen Folge der Erscheinungen. Um es darzustellen, bestimmt der menschliche Geist das empirisch Wankende, schließt das Zufällige aus, sondert das Unreine, entwickelt das Verworrene, ja entdeckt das Unbekannte.
>
> *(LA I 3 p. 308)*

22 Erster Entwurf einer allgemeinen Einleitung in die vergleichende Anatomie, ausgehend von der Osteologie (LA I 9, pp. 119–115).

23

> 1) in Beziehung aufs Auge
> physiologische
> 2) in Beziehung auf Licht und Finsternis
> physische.
>
> *(LA I 3, p. 320)*

24 "in Beziehung auf die Körper, an denen sie erscheint [...] Unter chemischen Farben verstehe ich Pigmente" (LA I 3, pp. 319–320).

25 Douglas Miller also notes that a "poetic treatment of the theme is found in [Goethe's poem] 'Reunion'" (Miller in Goethe 1995, p. 333).

26 Zwei Forderungen entstehen in uns bei Betrachtung der Naturerscheinungen: die Erscheinungen selbst vollständig kennenzulernen, und uns dieselben durch Nachdenken anzueignen. Zur Vollständigkeit führt die Ordnung, die Ordnung fordert Methode, und die Methode erleichtert die Vorstellung. Wenn wir einen Gegenstand in allen seinen Teilen übersehen, recht fassen und ihn im Geiste wieder hervorbringen können; so dürfen wir sagen, daß wir ihn im eigentlichen und im höhern Sinne anschauen, daß er uns angehöre, daß wir darüber eine gewisse Herrschaft erlangen. Und so führt uns das Besondere immer zum Allgemeinen, das Allgemeine zum Besondern. Beide wirken bei jeder Betrachtung, bei jedem Vortrag durcheinander. (LA I 3, p. 416. This essay is known by two titles *Introduction to the Lectures on Physics* [*Physikalische Vorträge schematisiert: Einleitung*] (in LA I 11, pp. 55–56 and [*Polarität*] in WA II 11, pp. 164–166; alternate translation also titled *Polarity* by Douglas Miller in Goethe 1995, pp. 155–156; translation also in Matthaei in Goethe 1971, p. 68).

27 "Von der vierten Art, welche bloß auf Anschauungen ruht, kann in der Sprache nichts vorkommen" (LA I 11, p. 57; translation taken from Miller in Goethe 1995, p. 27).

28 Es hat sich daher auch in dem wissenschaftlichen Menschen zu allen Zeiten ein Trieb hervorgetan die lebendigen Bildungen als solche zu erkennen, ihre äußern sichtbaren, reiflichen Teile in Zusammenhänge zu erfassen, sie als Andeutungen des Innern aufzunehmen und so das Ganze in der Anschauung gewissermaßen zu beherrschen. (LA I 9, p. 7; translation Miller in Goethe 1995, p. 63)

29 (Boyle 2003, p. 97). Boyle cites LA II 3, p. 404 as the basis for this argument, which states "... and challenged the objection against Newton" ["... und den Widerspruch gegen Newton herausforderte"], which is hardly "hostile".

30 Die Kraft, Farben hervorzubringen, ist von der Refraktion unabhängig (LA I 3, p. 447).

31 In the final chapter of his Historical Part, *Confessions of the Author* [*Konfession der Verfassers*], Goethe recalls the first time he looked through a prism

> I was in a totally white room; I held the prism before my eyes, I expected, remembering New-

tonian, that the entire white wall would be coloured in different shades [...] I was surprised that the white wall viewed through the prism remained white after all [Eben befand ich mich in einem völlig geweißten Zimmer; ich erwartete, als ich das Prisma vor die Augen nahm, eingedenk der Newtonischen Theorie, die ganze weiße Wand nach verschiedenen Stufen gefärbt ... Aber wie verwundert war ich, als die durchs Prisma angeschaute weiße Wand nach wie vor weiß blieb]

(LA I 6, pp. 419–420).

32 Es ist hier ein Dilemma, fast dem ähnlich, ob sich die Sonne um die Erde oder die Erde um die Sonne dreht, aus beiden lassen sich die Phänomene mehr oder weniger erklären. Nimmt man das Newtonische an und fängt von dem farbigen Bild an, so erklärt sich der weiße Raum sehr schwer. Fängt man von den farbigen Rändern an, so nimmt man den weißen Raum für das, was er ist, für unverändertes Licht und das Grün aus Gelb und Blau, das Purpur aus Blau- und Gelbrot, erklärt sich durch Mischung sehr schön. (LA I 3, p. 447)

33 I refer the reader to the excellent account by Sepper (1988).

34 LA II 3, p. 404.

35 To be fair, Newton did not know about wavelengths.

36 Goethe tells us so. The term Urphenomenon is used once in the entire Polemic Part, namely "the archetypical phenomena have been dealt with in detail in the *Outline* [the Didactic Part]" (Goethe 2016, p. 226) ["673. Uns sind diese Urphänomene aus dem Entwurf umständlich bekannt ..." (LA I 5, p. 190)].

37 "Without reading the polemical part of the *Farbenlehre* one cannot get a real feeling of the psychological abyss of Goethe's self-deceit" (Jaki 1969, p. 198). See also Duck in Goethe (2016, pp. xxiii–xxvii).

38 *Dedication, Preface and Didactic Part* [*Widmung, Vorwort und Didaktischer Teil*] (LA I 4).

39 *Exposing Newton's Theory* [*Enthüllung der Theorie Newtons*] (LA I 5). Eastlake (in Goethe 1840, p. xxxiv) called this section "controversial".

40 *Materials for the History of Colour Theory* [*Materialien zur Geschichte der Farbenlehre*] (LA I 6).

41 "... the present edition translation is confined to the first of these [Didactic Part], with such extracts from the other two as seemed necessary, in fairness to the author, to explain some of his statements" (Eastlake in Goethe 1840, p. xxxiv); and, "In Matthaei's [1971] edition the historical part of the *Farbenlehre* is completely omitted (save for a few passages from the 'Confessions of the Author,' a chapter that in the original edition concludes the historical part)" (Barasch 1974, p. 619).

42 "Deswegen finden wir, daß die Menschen lieber durch eine allgemeine theoretische Ansicht, durch irgend eine Erklärungsart die Phänomene bei Seite bringen, anstatt sich die Mühe zu geben, das Einzelne kennen zu lernen und ein Ganzes zu erbauen" (LA I 4, p. 17; translation by Douglas Miller in Goethe 1995, p. 163).

43 "Jene sind unaufhaltsam flüchtig, die andern vorübergehend, aber allenfalls verweilend, die letzten festzuhalten bis zur spätesten Dauer" (LA I 4, p. 20; translation by Douglas Miller in Goethe 1995, p. 165).

44 "Diese Farben, welche wir billig obenan setzen, weil sie dem Subjekt, weil sie dem Auge, teils völlig, teils größtens zugehören, diese Farben ..." (LA I 4, p. 25, §1; translation by Douglas Miller in Goethe 1995, p. 168).

45 Wenn wir die Augen innerhalb eines ganz finstern Raums offen halten, so wird uns ein gewisser Mangel empfindbar. Das Organ ist sich selbst überlassen, es zieht sich in sich selbst zurück, ihm fehlt jene reizende befriedigende Berührung, durch die es mit der äußern Welt verbunden und zum Ganzen wird [...] Wenden wir das Auge gegen eine stark beleuchtete weiße Fläche, so wird es geblendet und für eine Zeit lang unfähig, mäßig

beleuchtete Gegenstände zu unterscheiden. (LA I 4, p. 26, §6-7; translation by Douglas Miller in Goethe 1995, p. 168)

46 Die Netzhaut befindet sich bei dem, was wir sehen heißen, zu gleicher Zeit in verschiedenen, ja in entgegengesetzten Zuständen. Das höchste nicht blendende Helle wirkt neben dem völlig Dunkeln. Zugleich werden wir alle Mittelstufen des Helldunkeln und alle Farbenbestimmungen gewahr. (LA I 4, p. 27, §13; translation by Douglas Miller in Goethe 1995, p. 169)

47 "Wir wollen gedachte Elemente der sichtbaren Welt nach und nach betrachten und bemerken, wie sich das Organ gegen dieselben verhalte, und zu diesem Zweck die einfachsten Bilder vornehmen" (LA I 4, p. 27, §14; translation by Douglas Miller in Goethe 1995, p. 169).

48 "Wir blicken von einem Gegenstand auf den andern, die Sukzession der Bilder scheint uns rein, wir werden nicht gewahr, daß sich von dem vorhergehenden etwas ins nachfolgende hinüberschleicht" (LA I 4, p. 29, §19; translation by Douglas Miller in Goethe 1995, p. 170).

49 Man halte ein kleines Stück lebhaft farbigen Papiers oder seidnen Zeuges vor eine mäßig erleuchtete weiße Tafel, schaue unverwandt auf die kleine farbige Fläche und hebe sie, ohne das Auge zu verrücken, nach einiger Zeit hinweg, so wird das Spektrum einer andern Farbe auf der weißen Tafel zu sehen sein. Man kann auch das farbige Papier an seinem Orte lassen und mit dem Auge auf einen andern Fleck der weißen Tafel hinblicken, so wird jene farbige Erscheinung sich auch dort sehen lassen: denn sie entspringt aus einem Bilde, das nunmehr dem Auge angehört. (LA I 4, p. 37, §49; translation by Douglas Miller in Goethe 1995, p. 175)

Again "image" is a better term than "form", as it is an image that we retain when we undertake such an experiment.

50 Um in der Kürze zu bemerken, welche Farben denn eigentlich durch diesen Gegensatz her- vorgerufen werden, bediene man sich des illuminierten Farbenkreises unserer Tafeln, der überhaupt naturgemäß eingerichtet ist und auch hier seine guten Dienste leistet, indem die in demselben diametral einander entgegengesetzten Farben diejenigen sind, welche sich im Auge wechselweise fordern. So fordert Gelb das Violette, Orange das Blaue, Purpur das Grüne, und umgekehrt. So fordern sich alle Abstufungen wechselweise, die einfachere Farbe fordert die zusammengesetztere, und umgekehrt. (LA I 4, p. 38, §50; translation by Douglas Miller in Goethe 1995, pp. 175–176)

51 See Sepper (1988, pp. 12–13).

52 "Das Auge verlangt dabei ganz eigentlich Totalität und schließt in sich selbst den Farbenkreis ab" (LA I 4, p. 42, §60; translation by Douglas Miller in Goethe 1995, p. 178).

53 Man nennt dioptrische Farben diejenigen, zu deren Entstehung ein farbloses Mittel gefordert wird, dergestalt daß Licht und Finsternis hindurchwirken, entweder aufs Auge, oder auf ent- gegenstehende Flächen. Es wird also gefordert, daß Mittel durchsichtig oder wenigstens bis auf einen gewissen Grad durchscheinend sei. (LA I 4, p. 63, §143; translation by Douglas Miller in Goethe 1995, p. 190)

54 "Urphänomen" (LA I 4, p. 70, §174).

55 Das, was wir in der Erfahrung gewahr werden, sind meistens nur Fälle, welche sich mit einiger Aufmerksamkeit unter allgemeine empirische Rubriken bringen lassen. Diese subordinieren sich abermals unter wissenschaftliche Rubriken, welche weiter hinaufdeuten, wobei uns gewisse unerläßliche Bedingungen des Erscheinenden näher bekannt werden. Von nun an fügt sich alles nach und nach unter höhere Regeln und Gesetze, die sich aber nicht durch Worte und Hypothesen dem Verstande, sondern gleichfalls durch Phänomene dem Anschauen offenbaren. Wir nennen sie Urphänomene, weil nichts in der Erscheinung über ihnen liegt, sie aber dagegen völlig geeignet sind, daß man stufenweise, wie wir vorhin hinaufgestiegen, von ihnen herab bis zu dem gemeinsten Falle der

täglichen Erfahrung niedersteigen kann. (LA I 4, p. 71, §175; translation by Douglas Miller in Goethe 1995, pp. 194–195)

56 Wäre denn aber auch ein solches Urphänomen gefunden, so bleibt immer noch das Übel, daß man es nicht als ein solches anerkennen will, daß wir hinter ihm und über ihm noch etwas Weiteres aufsuchen, da wir doch hier die Grenze des Schauens eingestehen sollten. Der Naturforscher lasse die Urphänomene in ihrer ewigen Ruhe und Herrlichkeit dastehen, der Philosoph nehme sie in seine Region auf, und er wird finden, dass ihm nicht in einzelnen Fällen, allgemeinen Rubriken, Meinungen und Hypothesen, sondern im Grund- und Urphänomen ein würdiger Stoff zu weiterer Behandlung und Bearbeitung überliefert werde. (LA I 4, pp. 71–72, §177; translation by Douglas Miller in Goethe 1995, p. 195)

57 "Wenn wir von katoptrischen Farben sprechen, so deuten wir damit an, dass uns Farben bekannt sind, welche bei Gelegenheit einer Spiegelung erscheinen" (LA I 4, p. 124, §366; translation by Douglas Miller in Goethe 1995, p. 225).

58 "Die paroptischen Farben werden wir diejenigen nennen, welche entstehen, wenn das Licht an einem undurchsichtigen farblosen Körper herstrahlt" (LA I 4, pp. 129–130, §388; translation by Douglas Miller in Goethe 1995, p. 228).

59 "Sie entspringen durch verschiedene Veranlassungen auf der Oberfläche eines farblosen Körpers, ursprünglich, ohne Mitteilung, Färbe, Taufe ..." (LA I 4, pp. 139–140, §430; translation by Douglas Miller in Goethe 1995, p. 233).

60 "Allgemeine Ansichten nach Innen" (LA I 4, p. 203; translation by Douglas Miller in Goethe 1995, p. 266).

61 Wenn nun zwei aus derselben Quelle entspringende entgegengesetzte Phänomene, indem man sie zusammen bringt, sich nicht aufheben, sondern sich zu einem dritten angenehm Bemerkbaren verbinden, so ist dies schon ein Phänomen, das auf Übereinstimmung hindeutet. Das Vollkommnere ist noch zurück. (LA I 4, p. 206, §698; translation by Douglas Miller in Goethe 1995, p. 268)

62 "Die Farbenlehre besonders hat sehr viel gelitten, und ihre Fortschritte sind äußerst gehindert worden, daß man sie mit der übrigen Optik, welche der Messkunst nicht entbehren kann, vermengte, da sie doch eigentlich von jener ganz abgesondert betrachtet werden kann" (LA I 4, p. 212, §725; translation by Douglas Miller in Goethe 1995, p. 272).

63 Der Zustand, in welchem sich die allgemeine Physik gegenwärtig befindet, scheint auch unserer Arbeit besonders günstig, indem die Naturlehre durch rastlose, mannigfaltige Behandlung sich nach und nach zu einer solchen Höhe erhoben hat, daß es nicht unmöglich scheint, die grenzenlose Empirie an einen methodischen Mittelpunkt heranzuziehen. (LA I 4, p. 216, §738; translation by Douglas Miller in Goethe 1995, p. 274)

64 So haben auch wir seit langer Zeit den Ausdruck der Polarität in die Farbenlehre einzuführen gewünscht; mit welchem Rechte und in welchem Sinne, mag die gegenwärtige Arbeit ausweisen. Vielleicht finden wir künftig Raum, durch eine solche Behandlung und Symbolik, welche ihr Anschauen jederzeit mit sich führen müsste, die elementaren Naturphänomene nach unserer Weise aneinander zu knüpfen, und dadurch dasjenige deutlicher zu machen, was hier nur im allgemeinen und vielleicht nicht bestimmt genug ausgesprochen worden. (LA I 4, p. 223, §757; translation by Douglas Miller in Goethe 1995, p. 278)

65 "[Polarität]" in WA II 11, pp. 164–166. The original title in WA is in square brackets.

66 The title is given in WA II 11, pp. 164–166. Goethe returns to polarity in 1820 in his Entopic Colours [Entopische Farben LA I 8, pp. 94–138], "When we attribute polarity to entoptic phenomena, we do so in the sense in which we have endeavored to develop all chromagenesis in our theory of color" ["Wenn wir den entoptischen Phänomenen Polarität zuschreiben, so geschieht es in dem Sinne, wie wir in unserer Farbenlehre alle Chromagenesie zu entwickeln bemüht gewesen" (LA I 8, p. 99)].

67 Goethe also considered the magnet as an Urphenomenon "... this is clear the instant we say it. Thus it also comes to symbolise all else for which no words or names must be sought" (Goethe 1995, p. 303; also see Goethe's experiment with a magnet LA I 11, pp. 46–48). Goethe is not clear about *how* the magnet serves as an abstraction that we can

only see in our mind's eye. While Goethe is particular that the Urphenomenon is experience in our mind, a magnet is clearly a physical object. Could Goethe be referring to magnetism in general, or was it just a brain fart?

68 "Sinnlich-sittliche Wirkung der Farbe" (LA I 4, p. 224, §758; translation by Douglas Miller in Goethe 1995, p. 278).

69 "Indem ich diese Arbeit, welche mich lange genug beschäftigt, doch zuletzt nur als Entwurf gleichsam aus dem Stegreife herauszugeben im Falle bin ..." (LA I 4, p. 265; translation by Douglas Miller in Goethe 1995, p. 297).

70 In the chapter of the Historical Part, *Confession of the Author* [*Konfession der Verfassers*], Goethe tells us of his introduction into colour theory and how "In this way, the current outline of a theory of colours was created" ["Auf diese Weise ist der gegenwärtige Entwurf einer Farbenlehre entstanden"] (LA I 6, p. 427).

71 "... die sich auf der Oberfläche der Körper unter verschiedenen Bedingungen flüchtig oder bleibend erweisen" (LA I 8, p. 95).

72 Wenn wir den entoptischen Phänomenen Polarität zuschreiben, so geschieht es in dem Sinne, wie wir in unserer Farbenlehre alle Chromagenesie zu entwickeln bemüht gewesen. Finsternis und Licht stehen einander uranfänglich entgegen, eins dem andern ewig fremd; nur die Materie, die in und zwischen beide sich stellt, hat, wenn sie körperhaft undurchsichtig ist, eine beleuchtete und eine finstere Seite, bei schwachem Gegenlicht aber erzeugt sich erst der Schatten. Ist die Materie durchscheinend, so entwickelt sich in ihr im Helldunklen, Trüben in Bezug aufs Auge das, was wir Farbe nennen.

Diese, sowie Hell und Dunkel, manifestiert sich überhaupt in polaren Gegensätzen. Sie können auf gehoben, neutralisiert, indifferenziert werden, so daß beide zu verschwinden scheinen; aber sie lassen sich auch umkehren, und diese Umwendung ist allgemein bei jeder Polarität die zarteste Sache von der Welt. Durch die mindeste Bedingung kann das Plus in Minus, das Minus in Plus verwandelt werden. Dasselbe gilt also auch von den entoptischen Erscheinungen. Durch den geringsten Anlaß wird das weiße Kreuz in das schwarze, das schwarze in das weiße verwandelt und die begleitenden Farben gleichfalls in ihre geforderten Gegensätze umgekehrt. Dieses aber auseinander zulegen, ist gegenwärtig unsere Pflicht. Man lasse den Hauptbegriff nicht los, und man wird bei aller Veränderlichkeit die Grunderscheinung immer wieder finden. (LA I 8, p. 99)

73 Other authors have used the example of *Ranunculus acris* to show this gradation between forms (Brady 1987, p. 276, fig. 4).

74 Das Höchste, wozu der Mensch gelangen kann [...] ist das Erstaunen, und wenn ihn das Urphänomen in Erstaunen setzt, so sei er zufrieden; ein Höheres kann es ihm nicht gewähren, und ein Weiteres soll er nicht dahinter suchen; hier ist die Grenze. Aber den Menschen ist der Anblick eines Urphänomens gewöhnlich noch nicht genug, sie denken, es müsse noch weiter gehen, und sie sind den Kindern ähnlich, die, wenn sie in einen Spiegel geguckt, ihn sogleich umwenden, um zu sehen, was auf der anderen Seite ist. (Gespräche II, pp. 50–51)

75 Helmholtz also damned Goethe's work on botany and zoology with faint praise,

> ... in natural history the decision which characteristics are to be looked upon as important for classification, and which as unimportant, what divisions of the animal and vegetable kingdoms are more natural than others, is really left to an instinct of this kind, acting without any strictly definable rule. And it is a very suggestive fact that it was an artist, Goethe, who gave the first impulse to the researches of comparative anatomy into the analogy of corresponding organs in different animals, and to the parallel theory of the metamorphosis of leaves in the vegetable kingdom ; and thus, in fact, really pointed out the direction which the science has followed ever since.
>
> *(Helmholtz 1885, p. 20)*

Again this shows the narrow-mindedness of physics.

REFERENCES

Barasch M (1974) Goethe's Color Theory by Ruppecht Matthaei. *The Art Bulletin*, 56: 617–619.

Boyle N (2003) *Goethe: The poet and the age* Volume 2. Cambridge University Press, Cambridge.

Brady RH (1987) Form and cause in Goethe's morphology. In *Goethe and the sciences: A reappraisal* (Eds. F Amrine, FJ Zucker, H Wheeler) pp. 257–300. D. Reidel, Dordrecht, Holland.

Frankel HR (2012) *The Continental Drift Controversy: Wegener and the early debate*. Cambridge University Press, Cambridge.

Goethe JW (1840) *Theory of colours* (Trans. C Eastlake). The MIT Press, Cambridge, MA.

Goethe JW (1971) *Goethe's colour theory* (Ed. R Matthaei) Studio Vista Ltd, London.

Goethe JW (1989) Italian journey. *Goethe: The collected works* (Ed. D Miller). Suhrkamp Publishers, New York.

Goethe JW (1995) Scientific studies (Vol. 12). (Ed. TZP Saine and JL Sammons). Princeton, New Jersey.

Goethe JW (2016) *Goethe's exposure of Newton's theory. A polemic on Newton's theory of light and colour* (Eds. M Duck and M Petry). Imperial College Press, London.

Helmholtz H (1885) *Popular lectures on scientific subjects* (Trans. E Atkinson). D. Appelton and Co., New York.

Jaki S (1969) Goethe and the physicists. *American Journal of Physics*, 37: 195–203.

Magnus R (1949) *Goethe as a scientist*. Henry Schuman, New York.

Newton I (1704) *Opticks: or, a treatise of the reflexions, refractions, inflexions and colours of light*. Royal Society, London.

O'Neil C (2016) *Weapons of math destruction*. Crown Publishers, New York.

Richards R (2002) *Romantic conception of life*. Chicago University Press, Chicago.

Sepper DL (1988) *Goethe Contra Newton: Polemics and the project for a new science of color*. Cambridge University Press.

Sepper DL (2009) Goethe, Newton, and the imagination of modern science. *Revue Internationale de Philosophie*, 63: 261–277.

Sullivan HI (2017) Goethe's colors: Revolutionary optics and the anthropocene. *Eighteenth-Century Studies*, 51: 115–124.

Virchow R (2008) *Virchow's Eulogies*. Birkhäuser, Berlin.

Wilkins JS, Ebach MC (2014) *Nature of Classification*. Springer, New York.

4 Goethe on Botany

Goethe's writings on botany are not extensive. Far from it, they only cover *The Metamorphosis of Plants*[1] (hereon *Metamorphose*), and several short essays on plant physiology, personal accounts, book reviews and poems. Botany is Goethe's smallest contribution to his scientific work, yet it has received the greatest attention. *Metamorphose*, for example, was translated into French twice during Goethe's lifetime: by Frédéric Gingins-Lassaraz in 1829, and Frédéric Jacob Soret in 1831. The first English version was translated by Emily Cox in 1863, and others by Agnes Arber in 1946, Bertha Mueller in 1952 and Douglas Miller in 1988 [reprinted 1995]. Compare this to Goethe's *Farbenlehre*, which has never been translated in full, or *Beiträge*, which has only been translated once by Herb Aach (in Matthaei in Goethe 1971).

The rest of Goethe's writings of botany are often overlooked, with the exception of the *Urpflanze* or the "primordial plant", an idea he only mentions in personal letters published in his *Italian Journey*, but rarely in any of his scientific writings.

URPFLANZE

Nothing in Goethe's scientific writings seemingly captures the mind more than the Urpflanze. Much has been written about it and yet Goethe never really defines or discusses the Urpflanze in his scientific writings, choosing instead to leave the bulk of it in his *Italian Journey*,

> … perhaps all plant forms can be derived from one plant. Only in this way would it be possible to determine genera and species, which, it seems to me, has therefore been done very arbitrarily. My botanical philosophy remains stuck on this point, and I do not yet see how to proceed. The depth and breadth of the problem seem equally great to me.[2]
>
> *(Padua, 27 September 1786)*

> Please tell Herder that I shall soon have figured out the [Urpflanze]. Only I fear that no one will be willing to recognise the rest of the plant kingdom in it.[3]
>
> *(Naples, 25 March 1787)*

> I was taken again by my old fanciful idea: might I not discover the [Urpflanze] amid this multitude? Such a thing must exist, after all! How else would I recognise this or that form as being a plant, if they were not all constructed according to one [pattern]?[4]
>
> *(Palermo, 17 April 1787)*

DOI: 10.1201/9781003275343-4

> The [Urpflanze] is turning out be the most marvellous creation in the world, and nature itself will envy me because of it. With this model and key to it an infinite number of plants can be invented, which must be logical, that is, if they do not exist, they could exist, and are not mere artistic or poetic shadows and semblances, but have an inner truth and necessity. The same law will be applicable to every other living thing.[5]
>
> *(Naples, 17 May 1787)*

Boyle (1991, p. 501) notes,

> … it will take Goethe another three years to formulate his botanical principles, and by then the concept of the primal plant [Urpflanze] will be practically forgotten.

The Urpflanze is often confused with the "symbolic plant" that is featured in *Fortunate Encounter*. After attending a lecture, Schiller and Goethe fell into a conversation,

> [w]e reached [Schiller's] house, and our conversation drew me in. There I gave an enthusiastic description of the metamorphosis of plants, and with a few characteristic strokes of the pen I caused a symbolic plant to spring up before his eyes.[6]

The symbolic plant does not refer to the Urpflanze, an idea Goethe had by this time seemingly abandoned. Rather the drawing would have been "a representation of the different stages of metamorphosis" (see Boyle 2000, pp. 837–838, footnote 223), or a hypothesis of a whole plant.

The Urpflanze returns briefly in 1816, in a letter to Christian Gottfried Daniel Nees von Esenbeck (1776–1858)

> In the diary of my Italian trip, which is now being printed, you will notice, not without a smile, the strange ways in which I have followed the vegetative transformation. At that time I was looking for the [Urpflanze], unconscious that I was looking for the idea and the concept by which we could develop it.[7]

Again, the Urpflanze appears as a hypothesis from which the Urphenomenon could be discovered through Anschauung. The Urpflanze reappears, undefined, in Goethe's retelling of the history of his scientific studies.

> As they can now be gathered under a single concept, it gradually became clearer to me that the [Anschauung] could be enlivened in an even higher way: a requirement that then occurred to me under the sensory form of a super-sensible [Urpflanze]. I followed all the forms as they appeared to me, in their transformations, and thus, at the final destination of my journey, in Sicily, the original identity of all plant parts became completely clear to me, and I sought to pursue this everywhere and regain my awareness.[8]

These memories of Goethe's scientific thoughts were most likely written in the early 1800s, possibly around the time he was developing his Urphenomenon, before *Farbenlehre* and after his *Zur Morphologie*.

Goethe's Urpflanze is an idea of the whole, an abstract hypothesis that, through further Anschauung, can lead to the Urphenomenon.[9] The letter to von Esenbeck might provide a clue that gives a hint the Urpflanze was a precursor to the *Metamorphosis of*

Plants.[10] The Urpflanze represents the hypothesis of a whole or unity, of a plant of the entire taxon, be it a genus, family or order. For the Urpflanze to become a reality the observer needs to experience many plants via Anschauung. The reason why Goethe fails to refer to the Urpflanze in his later work is because it was superseded, in his methodology, by the Urphenomenon. In other words, the Urpflanze was a hypothesis that is "tested" by Anschauung and revealed through the Urphenomenon. The lack of any definition or discussion of the Urpflanze by Goethe hasn't stopped generations of historians and scientists from writing about it.[11]

On 8 March 1831, in a letter to his archivist Friedrich Theodor David Kräuter (1790–1856), Goethe wrote "I would like to know some biographical information about the French botanist and plant illustrator Turpin".[12] Goethe was interested in the French botanical illustrator Pierre Jean François Turpin (1775–1840) because of his accurate and detailed drawings of plants he drew for the French edition of *Metamorphose*.

Goethe was quite taken with Turpin's drawings in *Mémoires du Muséum d'histoire naturelle*, and by the illustrator's epigram in which he quotes himself,

> The general organisation of a living being and that of its organs in particular can only be explained by following step by step the successive development of this being, from the first moment of its apparent formation until that of its death.[13]

Turpin could be describing the Urphenomenon, and perhaps this is what caught Goethe's attention.[14] Moreover, Turpin illustrates this development by the humble potato (*Solanum tuberosum*; Figure 4.1). The figure captured Goethe's imagination,

> The excellent Turpin has had many opportunities to apply the highest principles that can be applied to plants, not only through scientific observation but also through artistic reproduction, and would therefore be of great service to this field if he were to seriously apply his skills to the purpose of a pictorial representation of plant metamorphosis.[15]

The translation of this paragraph in Goethe's *Oeuvre* in 1837 by Charles Frédéric Martins (1806–1889) has a footnote attached,

> Goethe's wish was fulfilled since Mr. Turpin was kind enough to adorn this book with plates III, IV, and V which are the visible reproduction of his ideas, and those of the illustrious poet, whose glorious legacy he accepted.[16]

One of these plates is the famous "Urpflanze" drawing of Turpin[17] (Figure 4.2). Given that the plates were published in 1837, it seems unlikely that Goethe had seen Turpin's Urpflanze.

Turpin's Urpflanze also had a greater impact than any of Goethe's original drawings in *Metamorphose*. In 1848 Matthias Jacob Schleiden (1804–1881) dismissed Turpin's depiction of the Urpflanze,

> This much is certain, that the disagreeable, tasteless heaping up of a number of individually possible forms into such an actual vegetable monster as Turpin has perpetrated in his Atlas to Goethe's scientific works, is anything but what Goethe pictured to himself in his ideal plant.

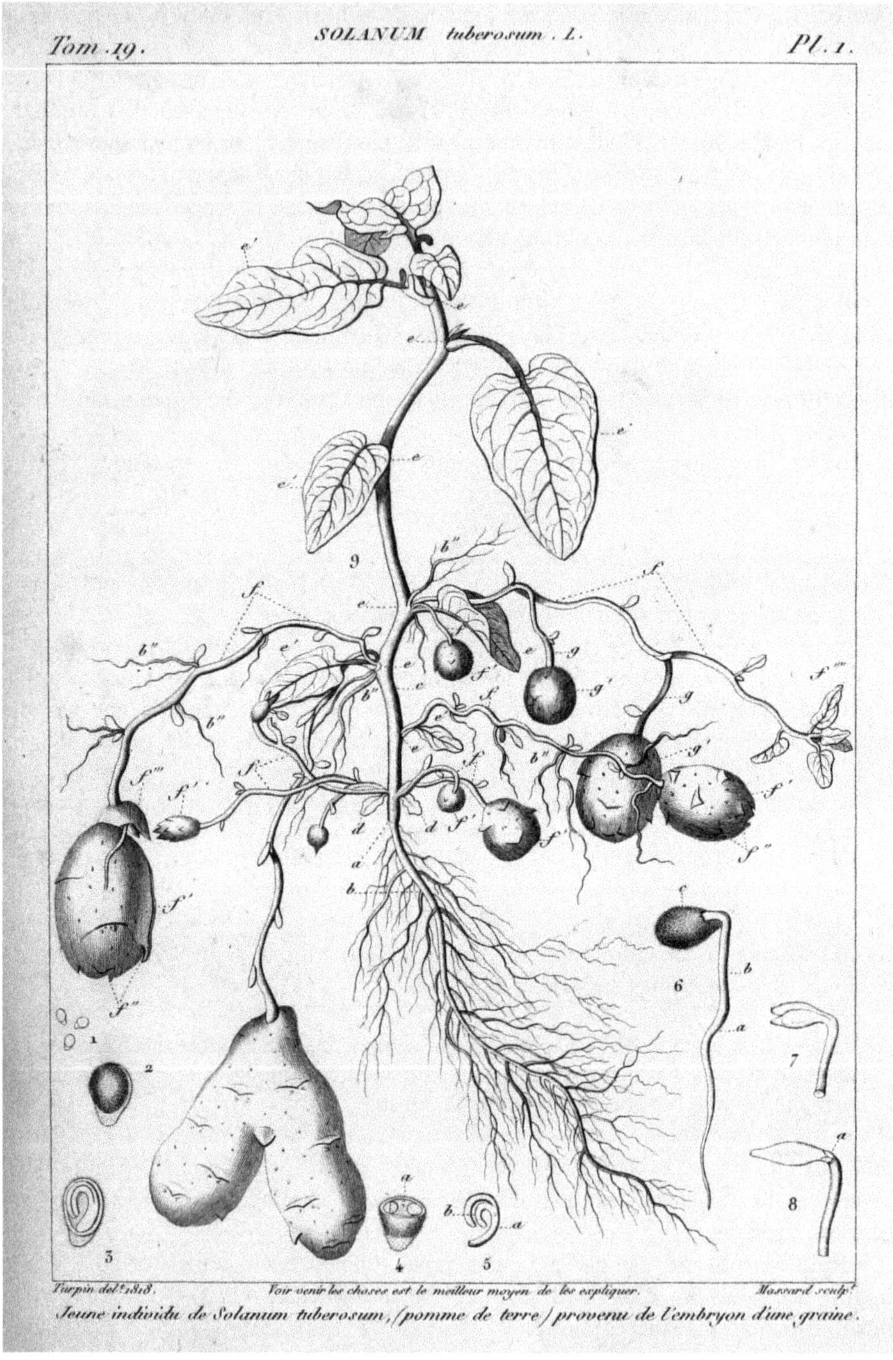

FIGURE 4.1 Turpin's drawings of *Solanum tuberosum* (Turpin 1830, Pl. 1). See text for details.

FIGURE 4.2 Turpin's famous Urpflanze of 1837. (Wikimedia.)

Schleiden even provided his own illustration[18] (Figure 4.3), so did Unger (1852), Sachs (1882), and Kerner von Marilaun (1883).[19]

Arber was perhaps the most pertinent in demystifying the Urpflanze. She too saw it as a precursor to the Urphenomenon, something which cannot be simply drawn in a single image (think, if you can, of a single illustration that shows the development of an egg, chick, pullet and hen). Arber tried to nip it in the bud (no pun intended),

> On his view the 'Urpflanze' could neither be described adequately in words, nor represented pictorially – an essential limitation which some of his followers unfortunately ignored.
>
> *(Arber 1949, p. 81)*

Yet the literature still embraces the Urpflanze even though, as Goethe points out in Turpin's 1830 illustration, it is, as Boyle notes, a series of images representing "the different stages of metamorphosis" rather than a single idealised plant.

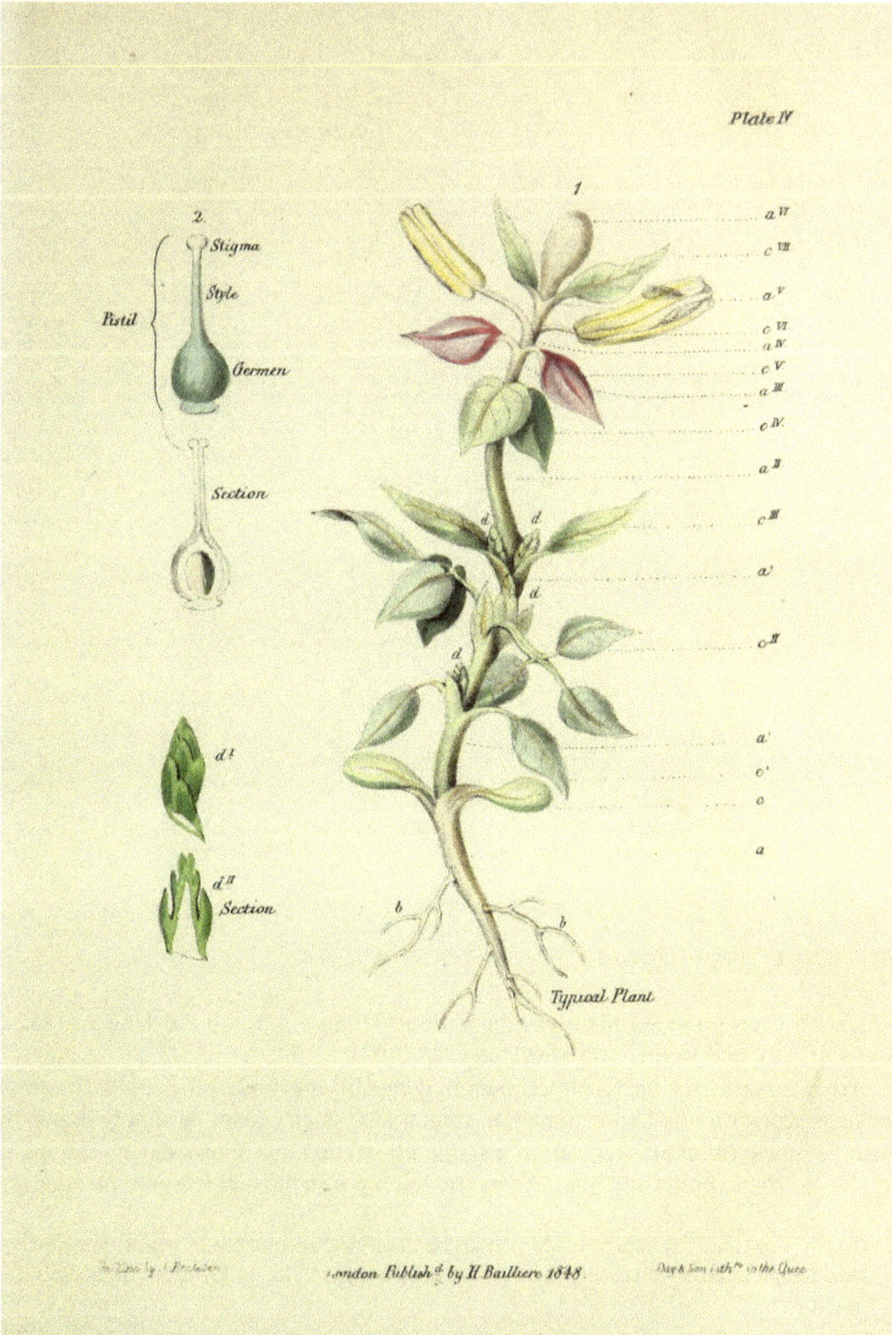

FIGURE 4.3 An abstraction of *Anagalis phoenicea* and *A. monelli*. See text for details.

THE METAMORPHOSIS OF PLANTS (1790)

Metamorphose was finished in January 1790, a year before the *Beiträge* were published, which possibly explains why they both are structured in the same way. Each work has a learning-by-doing approach; the *Beiträge* through doing experiments, so that by the end the experimenter/observer has undertaken a myriad of manifestations and reached the Urphenomenon. *Metamorphose* doesn't contain experiments per se; rather, they go through the life cycle of a plant via observation of each of the developmental stages of the organism. Over time, the observer can journey with Goethe and see an organ undergo a metamorphosis through a myriad of forms. The journey starts with seed leaves, the development of stem leaves from node to node, then transition to flowering, the formation of the calyx, the corolla, the stamens, the style and finally the fruits. *Metamorphose of plants* is just that, a progression of the same organ through a manifold of forms.

As with the *Beiträge*, Goethe also introduces a classification,

> This metamorphosis appears in three ways: regular, irregular and accident.[20]

Regular (or progressive) metamorphosis "can be seen to work step by step from the first seed leaves to the last formation of the fruit"[21] in the same way you would watch a seed grow into a fully mature plant over time. Irregular (or retrogressive) metamorphosis follows the same route as regular metamorphosis until there is "one or more steps backward", whereas accidental metamorphosis could "confuse our purpose [and] opportunities may arise elsewhere to speak monstrous but rather limited excrescences".[22] Unfortunately, Goethe didn't provide illustrations of these stages until versions that appeared after his death.[23]

Without illustrations or experiments, the *Metamorphose* lacks a subject that the author can observe, in fact Goethe is unapologetic about the absence of any visual guide,

> I have ventured to develop the present essay without reference to illustrations, although they might seem necessary in some respects.[24]

Goethe's lack of illustrations isn't sidestepping the issue of representation, but rather it encourages the reader/observer to grow the plant themselves, thereby adhering to the learning-by-doing approach. The same approach was used in the *Beiträge*; the cards only provided the reader/observer, armed with a glass prism, a *source* with which to see the phenomena, rather than showing illustrations of each phenomenon. The same can be said for the *Metamorphose*: a lack of images depicting the phenomenon only encourages the reader/observer to do the experiment themselves.

Metamorphose is in effect a field guide to the life of a plant from seed to mature plant,

> [s]ince we intend to observe the successive steps in plant growth, we will begin by directing our attention to the plant as it develops from the seed. At this stage we can easily and clearly recognise the parts belonging to it.[25]

More characteristics (i.e., plant structures) can be seen as the plant develops from a seed, moving from a generalised to a more specialised form. For example, compare the following two paragraphs,

> These first organs are known as *cotyledons* [...] They appear unformed, filled with a crude material, and as thick as they are broad. Their vessels are unrecognisable and scarcely distinguishable from the substance of the whole ...[26]
>
> Here we would obviously need general term to describe this organ [seed] which metamorphosed into such a variety of forms [Gestalten], a term descriptive of the standard against which to compare the various manifestations of its form [Gestalt].[27]

Goethe refers to two related concepts: first is wholeness in which a single general form manifests in multiple forms; the second is the Urphenomenon, in which the observer can comprehend the various manifestations mentally after having seen them as phenomena. Comparing manifestations of form is what taxonomists do when they observe organisms. What makes Goethe different from taxonomists of today and of the 1790s is that he explains his method.

Taxonomists also have "a term descriptive of the standard against which to compare the various manifestations of its form", namely, the "type". Unfortunately, this term has been formalised within the various Codes of Nomenclature to mean the name-bearing type, an individual specimen to which a scientific name is tied.[28] Goethe investigates the type [Typus] in his *Zur Morphologie* (see Chapter 5).

The concept for wholeness and the Urphenomenon is embryonic in the *Metamorphose*, but it's there in practise, just as it is in *Beiträge*. Goethe never followed up his *Metamorphose*[29] with a larger, more detailed work discussing his work "in a more precise and detailed way so that [his]point of view becomes clearer; perhaps then it will be more deserving of applause than at present".[30]

Goethe did, however, continue to develop his understanding of Anschauung and the Urphenomena and published *Attempts at a Botanical Method*[31] in 1795. In this work Goethe introduces his *Genetic Treatment*, in which he describes the Urphenomenon.

> When I see something that has emerged in front of me, so far as I can follow it, I will notice a series of steps that are unable to be seen together, rather I must vividly recall these in my memory as a certain ideal whole.[32]

Goethe correctly describes his Urphenomenon, a term not yet coined, which would later become his Pure Phenomenon in 1798 and finally the Urphenomenon in 1810. In 1795, Goethe also proposes the idea of the whole,

> At first I am inclined to think of certain stages, but since Nature does not make any jumps, I am finally forced to view the result of an uninterrupted activity as a whole, by discarding the individual without destroying the impression.[33]

Goethe gives examples where his genetic treatment has been used,

> The teaching [Lehre] of the *Metamorphosis of Plants* in its entire meaning.[34]

Goethe's *Attempt at a Botanical Method* includes outlines of *Organic Unity* [*Organische Einhalt*], *Organic Divisions* [*Organische Entzweiung*], *Physiological Concepts* [*Begriffe einer Physiologie*] and general meditations about the role of morphology in the sciences, which will be covered in the following chapters.

One sticking point that Goethe addressed later in life was his silence on the role of the root. The root isn't mentioned in his *Metamorphose* and only touched upon in other related writings on botany. Goethe wasn't fazed by this criticism, "What was I to do with a formation of filaments, fibres, bulbs and tubers [...] an endless variety with no Steigerung". What Goethe means is that the "filaments, fibres, bulbs and tubers" of plants rarely change over time and develop new characteristics. A potato tuber just grows and changes little, unlike a stem or flower. Where is the Urphenomenon in such a structure? Regardless, Goethe does comment on various other botanical phenomena (e.g., the leaf in relation to the root, spiral tendencies in plants and plant reproduction), with few of these notes pertaining to his method.

Goethe's Morphology wasn't restricted to the life sciences; it included geology[35] and colour theory. However, the aims underpinning this new science are explored more deeply in his zoological writings.

NOTES

1 While Goethe's work is known in English as *The Metamorphosis of Plants*, the title is best translated as *An Attempt to Explain the Metamorphosis of Plants* [*Versuch die Metamorphose der Pflanzen zu erklären*]. Of the seven English translations only two use the correct title (*sensu* Arber 1949; Goethe 2023).

2 Hier in dieser neu mir entgegentretenden Mannigfaltigkeit wird jener Gedanke immer lebendiger: daß man sich alle Pflanzengestalten vielleicht aus Einer entwickeln könne. Hierdurch würde es allein möglich werden, Geschlechter und Arten wahrhaft zu bestimmen, welches, wie mich dünkt, bisher sehr willkürlich geschieht. Auf diesem Punkte bin ich in meiner botanischen Philosophie stecken geblieben, und ich sehe noch nicht, wie ich mich entwirren will. Die Tiefe und Breite dieses Geschäfts scheint mir völlig gleich. (WA I 30, pp. 89–90; Translated by Robert Heitner in Goethe 1989, p. 214)

3 "Herdern bitte ich zu sagen, daß ich mit der Urpflanze bald zustande bin; nur fürchte ich, daß niemand die übrige Pflanzenwelt darin wird erkennen wollen" (WA I 31, p. 75; Translated by Robert Heitner in Goethe 1989, p. 181).

4 Im Angesicht so vielerlei neuen und erneuten Gebildes fiel mir die alte Grille wieder ein, ob ich nicht unter dieser Schar die Urpflanze entdecken könnte. Eine solche muß es denn doch geben! Woran würde ich sonst erkennen, daß dieses oder jenes Gebilde eine Pflanze sei, wenn sie nicht alle nach einem Muster gebildet wären? (WA I 31, pp. 147–148; Translated by Robert Heitner in Goethe 1989, p. 214)

Note that Miller choses to use the term "model" for the German word "Muster". "Muster" may also mean "pattern" or "design", which is closer to Goethe's intention, as the cotyledon Lehre looks for patterns common in all plants.

5 Die Urpflanze wird das wunderlichste Geschöpf von der Welt, um welches mich die Natur selbst beneiden soll. Mit diesem Modell und dem Schlüssel dazu kann man alsdann noch Pflanzen ins Unendliche erfinden, die konsequent sein müssen, das heißt: die, wenn sie auch nicht existieren, doch existieren könnten und nicht etwa malerische oder dichterische Schatten und Scheine sind, sondern eine innerliche Wahrheit und Notwendigkeit haben. Dasselbe Gesetz wird sich auf alles übrige Lebendige anwenden lassen. (WA I 31, p. 240; Translation by Robert Heitner in Goethe 1989, p. 299)

There is a discrepancy in the source literature over the date when the letter was written. In *Italian Journey* the letter is addressed to Herder and written in Naples 17 May 1787 (in WA I 31, p. 240). Another source claims the letter was written to Frau von Stein between 8 and 9 June 1787 (for more information, see LA II 9A, p. 365–366). The same passage is repeated in the third part of the *Italian Journey* under the title "Report" (Italian Journey, Naples 17 May 1787, Goethe 1989, p. 299; see also Ebach 2022, p. 62, footnote 30).

6 "Wir gelangten zu seinem Hause, das Gespräch lockte mich hinein; da trug ich die Metamorphose der Pflanzen lebhaft vor, und ließ, mit manchen charakteristischen Federstrichen, eine symbolische Pflanze vor seinen Augen entstehen" (LA I 9, p. 81; translation by Miller in Goethe 1995, p. 20).

7 In den Tagebüchern meiner Italienischen Reise, an welchen jetzt gedruckt wird, werden Sie, nicht ohne Lächeln, bemerken, auf welchen seltsamen Wegen ich der vegetativen

Umwandlung nachgegangen bin; ich suchte damals die Urpflanze, bewußtlos, daß ich die Idee, den Begriff suchte wonach wir sie uns ausbilden könnten. (WA IV 27, p. 144)

8 Wie sie sich nun unter einen Begriff sammeln lassen, so wurde mir nach und nach klar und klarer, daß die Anschauung noch auf eine höhere Weise belebt werden könnte: eine Forderung, die mir damals unter der sinnlichen Form einer übersinnlichen Urpflanze vorschwebte. Ich ging allen Gestalten, wie sie mir vorkamen, in ihren Veränderungen nach, und so leuchtete mir am letzten Ziel meiner Reise, in Sizilien, die ursprüngliche Identität aller Pflanzenteile vollkommen ein, und ich suchte diese nunmehr überall zu verfolgen und wieder gewahr zu werden. (LA I 10, p. 334)

9 The hypothesis that "everything is leaf" ["Alles ist Blat [sic]"] was written during Goethe's trip to Italy, around the same time when the Urpflanze was coined. The hypothesis was scribbled on several small sheets of paper. The fact that these small unpublished notes made it into the main literature on Goethe's botany is remarkable. The entire passage reads,

Hypothesis

Everything is leaf and [therefore] through this simplicity, the greatest diversity becomes possible.

The leaf has vessels that, when intertwined, produce another leaf, where one can form a crude image by intertwining two lines

Hypothese

Alles ist Blat [sic] und [dadurch] durch diese Einfachheit wird die größte Mannigfaltigkeit möglich.

Das Blatt hat Gefäße die in sich verschlungen wieder ein Blatt hervorbringen wo man ein krudes Bild durch Verschlingung zweier Linen sich Formiren kann. (WA II 7, p. 282)

10 Goethe provides evidence for this when he says "Please tell Herder that I shall soon have figured out the [Urpflanze]. Only I fear that no one will be willing to recognise the rest of the plant kingdom in it. My much-discussed [Lehre] of cotyledons is so highly refined that it would be difficult to go any further".

> Herdern bitte ich zu sagen, daß ich mit der Urpflanze bald zustande bin; nur fürchte ich, daß niemand die übrige Pflanzenwelt darin wird erkennen wollen. Meine famose Lehre von den Kotyledonen ist so sublimiert, daß man schwerlich wird weiter gehen können.
>
> *(WA I 31, p. 75; Translation by Robert Heitner in Goethe 1989, p. 181)*

The "Lehre of cotyledons" is most likely an early idea or draft of the *Metamorphosis of Plants*.

11 There are too many to list here, but recent works include Geulen (2012), Müller and Müller (2015) and Zemplén (2017).

12 "Von dem französischen Botaniker and Pflanzenzeichner Turpin wünsche einige biographischen Kenntnisse" (WA IV 48, p. 143).

13 L'organisation générale d'un être vivant et celle de ses organes en particulier ne peuvent s'expliquer qu'autant que l'on: suit pas à pas le développement successif de cet être, depuis le premier moment de sa formation apparente jusqu'à celui de sa mort. (Turpin 1820, pp. 15–16).

14 Perhaps the publication of 1830 led Goethe to ask Kräuter for Turpin's biography? Goethe was familiar with Turpin's monograph in the *Mémoires* from 5 December 1829 (see WA II, p. 162).

15 Das Höchste, was sich hier von auf Pflanzen anwenden läßt, hat der treffliche Turpin nicht allein durch wissenschaftliches Beschauen, sondern auch künstlerische Nachbildung zu bearbeiten vielfache Gelegenheit gehabt, und würde daher diesem Felde den größten Dienst leisten, wenn er seine Geschicklichkeit zum dem Zwecke einer bildlichen Darstellung der Pflanzenmetamorphose ernstlich hinleiten wollte. (LA I 10, p. 310; also "see Plates III, IV and V" ["Voy. pl. [Voyez les planches] III, IV et V]" in Goethe 1837a, p. 309; also see Goethe 1837b).

16

> Le voeu de Goethe a été accompli puisque M. Turpin a bien voulu orner ce livre des planches III, IV, et V, qui sont la reproduction visible de ses idées et de celles de l'illustre poëte, dont il accepté le legs glorieux.
>
> *(Martins in Goethe 1837a, p. 310, footnote *)*

17 Turpin's illustrations were published separately as *Atlas contenant deux planches d'anatomie comparée, trois de botanique et deux de géologie: accompagnées d'une explication raisonnée et d'une esquisse d'organographie végétale*, which contains 7 leaves of plates (see https://catalogue.nli.ie/Record/vtls000471587).

18 The accompanying plate (IV) will serve to represent such a plant. It may be regarded as an abstraction of a very simple and familiar plant, the *Anagalis phoenicea*, the large blue-flowered variety of which decorates our windows, under the name of *Anagalis monelli*. A close examination of this picture will render some of the more important morphological conceptions more readily intelligible. A mere glance even shows us the following characteristics: In the first place, we distinguish a continuous principal trunk (*a* to *a* vi), with various lateral appendages attached to it (*b, c* to *c* VI and *d*). Viewed more narrowly, these later exhibit some very striking differences, which are capable of being arranged into three classes (which, therefore, are distinguished by the three letters *b, c* and *d*). Continuing our investigation, we find the organs marked *d* (see Figs. *d* I, *d* II) to be also composed of a main trunk and lateral organs, and, having exactly the same characters as the plant itself in their subsequent development, they are mere repetitions of it, only distinguished by not being free at the lower extremity. (Schleiden, 1853, pp. 83–84)

19 You can compare them in Niklas and Kutschera (2017, fig. 1).

20 "Es zeigt sich uns diese Metamorphose auf dreierlei Art: regelmäßig, unregelmäßig und zufällig" (§5; LA I 9, p. 23; translation by Douglas Miller).

21 "… welche sich von den ersten Samenblättern bis zu letzen Ausbildung der Frucht …" (§6; LA I 9, p. 24; translated by Douglas Miller).

22 … denn sie ist es, welche sich von den ersten Samenblättern bis zur letzten Ausbildung der Frucht immer stufenweise wirksam bemerken läßt […] Denn wie in jenem Fall die Natur vorwärts zu dem großen Zwecke hineilt, tritt sie hier um eine oder einige Stufen rückwärts […] unsern Zweck verrücken könnte. Vielleicht findet sich an einem andern

Orte Gelegenheit, von diesen monströsen, und doch in gewisse Grenzen eingeschränkten Auswüchsen sprechen. (§6–8; LA I 9, p. 24; translated by Douglas Miller)

23 "Although Goethe had illustrations prepared [...], these were not published until after his death" Miller (in Goethe 1995, p. 327).

24 "Ich habe es gewagt, gegenwärtigen Versuch ohne Beziehung auf erläuternde Kupfer auszuarbeiten, die jedoch in manchem Betracht nötig scheinen möchten" (§9; LA I 9, p. 25, translated by Douglas Miller in Goethe 1995, p. 77).

25 Da wir die Stufenfolge des Pflanzen-Waschstums zu beobachten uns vorgenommen haben, so richten wir unsere Aufmerksamkeit sogleich in dem Augenblicke auf Pflanzen, da sich aus dem Samenkorn entwickelt. In dieser Epoche können wir die Teile, welche unmittelbar zu ihr gehörten, leicht und genau erkennen. (§10; LA I 9, p. 25, translated by Douglas Miller in Goethe 1995, p. 77)

26 Es sind diese ersten Organe unter dem Namen Kotyledonen bekannt [...] Sie erscheinen oft unförmlich, mit einer rohen Materie gleichsam ausgestopft, und eben so sauer in die Dicke als in die Breite ausgedehnt; ihre Gefäße sind unkenntlich und vor der Masse des Ganzen kann zu unterscheiden ... (§11–12; LA I 9, pp. 25–26, translated by Douglas Miller in Goethe 1995, p. 77, original emphasis)

27 Es versteht sich hier von selbst, daß wir ein allgemeines Wort haben müßten wodurch wir dieses in so verschiedene Gestalten metamorphosierte Organ bezeichnen, und alle Erscheinungen seiner Gestalt damit vergleichen könnten: gegenwärtig müssen wir uns damit begnügen, daß wir uns gewöhnen die Erscheinungen vorwärts und rückwärts gegeneinander zu halten. (§120; LA I 9, p. 60, translated by Douglas Miller in Goethe 1995, pp. 96–97)

28 The term "type" has also been misunderstood and demonised by protagonists of the Modern Synthesis. The less we speak of them, the better.

29 Only several smaller works followed, including an 11-paragraph-long *Metamorphosis of Plants: a second attempt*, in which he mostly refers the reader to his zoological work *Animal Form* [*Gestalt der Tiere*] (see LA I 1 pp. 64–67).

30 "... genauer und umständlicher abzuhandeln, um diese Vorstellungsart anschaulicher zu machen, und ihr einen allgemeinern Beifall zu erwerben, als sie vielleicht gegenwärtig nicht erwarten kann" (§123; LA I 9, p. 61, translated by Douglas Miller in Goethe 1995, p. 97).

31 "Versuche zur Methode der Botanik" (LA I 10, p. 129). The term "Versuche" can also be translated as "experiments". Since these are not strictly experiments, I have used the term "Attempts".

32

> Wenn ich eine entstanden Sache vor mir sehe weit ich ihn verfolgen kann, so werde ich eine Reihe Stufen gewahr die ich zwar nicht neben einander sehen kann sondern mir in der Erinnerung zu einem gewissen idealen Ganzen vergegenwärtigen muß.
>
> *(LA I 10, p. 131)*

33

> Erst bin ich geneigt mir gewisse Stufen zu denken, weil aber die Natur keinen Sprung macht, bin ich zuletzt genötigt mir die Folge einer ununterbrochenen Tätigkeit als ein Ganzes anzuschauen indem ich das Einzelne aufhebe ohne den Eindruck zu zerstören.
>
> *(LA I 10, p. 131)*

34 "Lehre von der Metamorphose der Pflanzen in ihrer ganzen Bedeutung" (LA I 10, p. 132).

35 Goethe had difficulty in incorporating geology into his Morphology; however, unlike Steigerwald (2002, p. 296, f. 10), I think it includes colour and sound.

REFERENCES

Arber A (1949) Goethe's botany. *Chronica Botanica*, 10: 63–126.
Boyle N (1991) *Goethe: The poet and the age* (Vol. 1). Oxford University Press, Oxford.
Boyle N (2000) *Goethe: The poet and the age* (Vol. 2). Oxford University Press, Oxford.
Ebach MC (2022) *Goethe in the Age of Artificial Intelligence*. Palgrave Macmillan, Singapore.
Geulen E (2012) *Urpflanze (und Goethes Hefte zur Morphologie).* (Eds. T Döring, M Ott) Urworte. Brill, Leiden, pp. 155–171.
Goethe JW (1837a) *Oeuvres d'histoire naturelle de Goethe: comprenant divers mémoires d'anatomie comparée de botanique et de géologie* (Translated and annotated by CF Martins). Abraham Cherbuliez, Paris.
Goethe JW (1837b) *Oeuvres d'histoire naturelle de Goethe: Atlas contenant deux planches d'anatomie comparée, trois de botanique et deux de géologie*. Abraham Cherbuliez, Paris.
Goethe JW (1863) Goethe's essay on the Metamorphosis of Plants (Trans. Emily M Cox). *Journal of Botany*, 1: 327–345, 360–374.
Goethe JW (1952) *Goethe's Botanical writings* (Trans. B Mueller), with an introduction by CJ Engard. Ox Box Press, Woodbridge, CT.
Goethe JW (1971) *Goethe's colour theory* (Ed. R Matthaei) Studio Vista Ltd, London.
Goethe JW (1989) *Italian Journey* (Trans. RR Heitner). Suhrkamp Publishers, New York.
Goethe JW (1995) Scientific studies (Vol. 12). *Goethe: The collected works* (Ed. D Miller). Princeton, New Jersey.
Goethe JW (2023) *An Attempt to Explain the Metamorphosis of Plants* (Trans. F Hollenbaum). Independently Published.
Kerner von Marilaun A (1883) *Pflanzenleben*, vol 1. Bibliographisches Institut, Leipzig
Müller WA, Müller WA (2015) *Goethes Urpflanze und ihre Metamorphose.* R-Evolution-des biologischen Weltbildes bei Goethe, Kant und ihren Zeitgenossen, 17–31.
Niklas KJ, Kutschera U (2017) From Goethe's plant archetype via Haeckel's biogenetic law to plant evo-devo 2016. *Theory in Biosciences*, 136: 49–57.
Sachs J (1882) *Vorlesungen über Pflanzenphysiologie*. W. Engelmann, Leipzig
Schleiden MJ (1853) *The Plant; a biography*. Hippolyte Bailliere, London.
Steigerwald J (2002) Goethe's Morphology: Urphänomene and Aesthetic Appraisal. *Journal of the History of Biology* 35:291–328.
Turpin PJF (1820) *Essai d'une iconographie élémentaire et philosophique des végétaux (in French).* Paris: Charles Louis Fleury Panckoucke.
Turpin PJF (1830) *Mémoire sur l'organisation intérieure et extérieure des tubercules du Solanum tuberosum et de l'Helianthus tuberosus, considérés comme une véritable tige souterraine, et sur un cas particulier de l'une de ces tiges*. Par A. Belin, Paris.
Unger F (1852) *Ideale Darstellung einer vollkommenen Pflanze*. Botanische Briefe, Wien
Zemplén GÁ (2017) *Structure and advancement in Goethe's morphology* (Ed. J Faflak) Marking Time, University of Toronto Press, Toronto, pp. 147–172.

5 Goethe on Zoology

The journal *Zur Morphologie*, compiled in 1817–1822, contains a large portion of Goethe's work on zoology and morphology, even though much of it was written before 1817. None of Goethe's zoological works were published as comprehensive works like books or treatises but are scattered in various documents and letters written between 1784 and the end of his life.

Goethe's first work was on the intermaxillary bone [*os intermaxillare*], which he wrote with Justus Ferdinand Christian Loder (1753–1832) between 1784 and 1786 and later published in 1820 in *Zur Morphologie*.

Titled *An Intermaxillary Bone is Present in the Upper Jaw of Man as Well as in Animals*[1] (1784–1786), Goethe describes the bone in anatomical detail, noting, "[w] ere observation not convincing evidence of things in Nature, I would be reluctant to appear with my claim that this bone is present in man as well".The bone, obvious in vertebrates but apparently absent in humans, had bolstered the late 18th-century view that man was separate from animals. Goethe, with the help of Loder, discovered it in human embryos and children as a fused bone, rather than two bones as in other vertebrates. Goethe's work resembles that of many late 18th-century and later 19th- and 20th-century studies, namely, it lacks a methodology or a methods section. Goethe refers to following "Camper's method exactly".

Petrus Camper (1722–1789) was one of the first practising comparative anatomists who developed a way of drawing through the concept of metamorphosis – think of *Australopithecus* turning into the various species of *Homo* (Figure 5.1).

Goethe was seemingly impressed with Camper's method,

> What a gulf the *os intermaxillare* of the turtle and the elephant, and yet an intermediate series of forms can be found to connect the two![2]

The *An Intermaxillary Bone* is unique as it predates Goethe's *Italian Journey*, and much of his later morphological musings even though they contain kernels of his later thought. Camper's metamorphosis starts with a prototype that can change its dimensions over time to resemble other forms. These are done on paper as a series of drawings in which measurements are taken, in a similar manner to the graphs drawn by D'Arcy Wentworth Thompson (1860–1948).[3] These graphs may be able to predict forms; however, they are not based on the experience of form, namely Anschauung. In this sense, these mathematical models are completely abstract and cannot be incorporated into the Urphenomenon until they are experienced via Anschauung. In any case, Goethe's disciplined observation and thinking required for Anschauung and the Urphenomenon didn't crystallise until a decade later during his work on zoology and colour.

While Goethe does refer to Camper's method and does provide images, they do not reveal the method Goethe was later to discover for himself – our ability to observe and recall, or Anschauung and the Urphenomenon. Yet Camper did provide Goethe

DOI: 10.1201/9781003275343-5

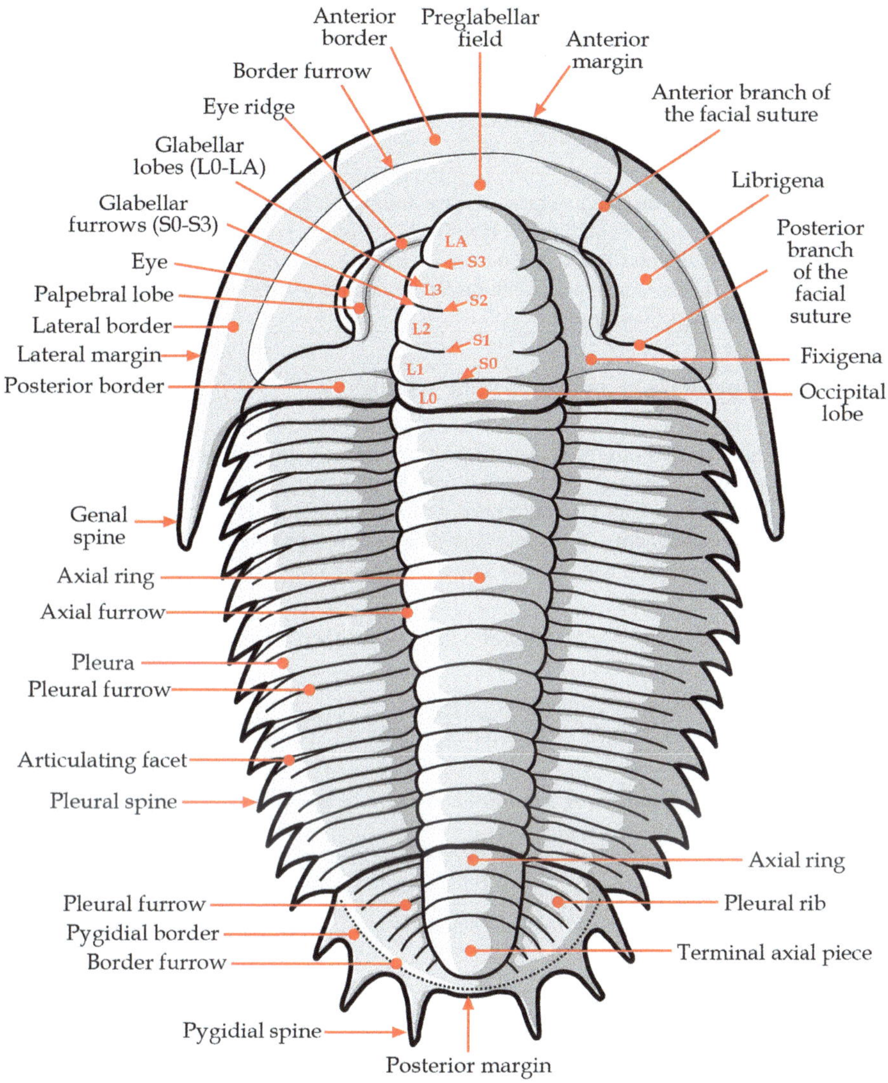

FIGURE 5.1 The anatomy of a trilobite, a Palaeozoic marine arthropod.

(Image by Sam Gon III used with permission.)

with the prototype or Urbild, which Goethe later developed into the *Typus* in his Morphology.

URBILD

The term Urbild (literally "primordial image") smacks of another case of the curse of Ur-[4] Unlike Urpflanze, Urbild was a term that was already in use in 18th-century

German. In fact, Urbild was first used by German dramatist Kaspar von Stieler (1691, p. 150) in his dictionary to denote an archetype. By the time Goethe used the term Urbild, its meaning included prototype, type or pattern.[5]

What did Goethe mean by Urbild, given the term was already in usage by the time he first used it in his *Italian Journey*?

> Marble is a strange material, which is why Apollo of Belvedere in [Urbilde] is so boundlessly pleasing, for the highest breath of the living, youth-free, eternally young being vanishes momentarily into the best plaster cast.[6]
>
> … it is also like happiness and wisdom, from which the [Urbilder] only hover before us, whose hems we at best touch.[7]

In 1796 the term shows up in his scientific writing.

> We would have gained this, then, by being able to assert without hesitation that all perfect organic natures, among which we see fish, amphibians, birds, mammals, and at the top of the last, man, are all formed according to an [Urbilde], which only differs more or less in its very permanent parts and is still daily developed and transformed through reproduction.[8]

Urbild also is used in the poem *Metamorphose der Tiere* (1806).

> All limbs are formed according to eternal laws,
> and the rarest form secretly preserves the [Urbild].[9]

The term also appears in the play *Pandora* (1810), in which the character Prometheus cries,

> Go peacefully! You won't find peace.
> Such a fortune has been given to humans and animals,
> After whose [Urbild] that I shaped better …[10]

Goethe never uses Urbild in his practise of science, meaning the term describes a general concept (i.e., archetype) rather than a specific activity (i.e., Anschauung). Goethe does however use the term Urtier in the introduction to his 1817 *Zur Morphologie*,

> I soon felt the necessity of establishing a *Typus* against which all mammals could be tested for similarities and differences, and just as I had previously sought the original plant, I now strove to find the [Urtier], which ultimately meant the concept, the idea of the animal.[11]

The "concept" or "idea" Goethe refers to here is the Urphenomenon. The Urtier is the animal in all its manifestations developed through Anschauung and which we can recall in our mind's eye. The Typus Goethe refers to is the template or outline that we can compare to all these manifestations and is still in use today in comparative biology and taxonomy.

TYPUS

The concept of the Typus[12] was introduced in *Outline for a General Introduction to Comparative Anatomy* (1795) to assist comparative anatomists in comparing the characteristics of form between different organisms. The Typus acts as a general outline or list of characteristics, from which all forms can be compared. The Typus is not a real organism but rather a general anatomical outline, a type of checklist (i.e., head, torso, forelimbs, hind-limbs). In this way a human arm and a bat wing can be compared as being two manifestations of the mammalian forelimb. In Goethe's words, "[e]mpirical observation must first teach us what parts are common to all animals, and how these parts differ".[13] Unlike the Urphenomenon, the Typus is a tool or aid that assists classification, that is, describing each organism.

> Hence, an anatomical [Typus] will be suggested here, a general picture containing the forms of all animals as potential, on which will guide us to an orderly description of each animal. As much as possible the [Typus] must be established physiologically. The mere idea of a [Typus] in general implies that no particular animal can be used as our point of comparison; the particular can never serve as a measure for the whole.[14]

Goethe divided up the animal body plan into three parts: the head (sensory part), the midsection (organs of life) and the rear section (digestion and reproduction). The auxiliary organs (fore and hind-limbs) are attached to either of these sections. Using this Typus (head, midsection, rear section and auxiliary organs) we have constructed a general form, a skeletal structure, from which we can compare other vertebrate organisms. Once such a Typus is formed for mammals, for instance, it would be almost impossible to miss the intermaxillary bone, because based on the skeletal structure, we would look for it and find a fused bone rather than two separate bones. Using the Typus means that the same structures that have different manifestations can be compared systematically.[15] Goethe classified the osteological Typus into its various parts, each one having an infinite manifestation of forms, that can be compared within an anatomical Typus,

> A. The Head
>   a. Ossa intermaxillaria
>   b. Ossa maxillae
>   c. Ossa palatinas
>   [...]
>
> B. The Torso
>   I. Spina dorsalis
>   a. Vertebrae colli
>   b. dorsi
>   c. lumborum
>   [...]
>
> C. Auxiliary organs
>   1. Maxilla inferior
>   2. Brachia
>   [...]
>
> *(LA I 9 130–132)*

Goethe's Typus is very much like the Italian architect Palladio's "Measures of the Tuscan order, as taught by Vitruvius" from his *I quattro libri dell'architettura*, namely,

A. Abacus
B. Ovolo, or Echinus
C. Collarino, or Frise of the Capitel
D. Astragal
E. The Body of the Column above
F. The Body of the Column below
G. Listella, or Cincture, or Annulet
H. Torus, or Tore
I. Orlo, or Plinth
J. Pedestal, or Stylobatum

(Palladio 1742, p. 15; Ebach 2022, p. 60; Table 5.1)

Palladio (1742) developed this system in order to compare the different forms of pillars. Goethe had bought a copy of *I quattro libri dell'architettura* during his Italian journey in 1786. Whether Goethe picked this method up from Palladio or whether Loder had shown him an entirely different method is unclear.

The Typus is clearly a universal way to assist comparative anatomy, through a rigorous anatomical nomenclature within an osteological classification. Although this method was new in Goethe's time, it is still used by comparative biologists (i.e., taxonomists) in the present day.

TABLE 5.1
The Animal Typus Compared to the Plant Typus via the "Measures of the Tuscan Order, as Taught by Vitruvius"

VI. The Osteological Type Organised in its parts (1817)	Measures of the Tuscan Order, as Taught by Vitruvius (Palladio 1742, p. 11)	Metamorphosis of Plants (1790)
A. The head	A. Abacus	I. Of the Seed Leaves
a. Ossa intermaxillaria	B. Ovolo, or Echinus	II. Development of the Stem Leaves from Node to Node
b. Ossa Maxillae superioris	C. Collarino, or Frise of the Capitel	III. Transition to Flowering
c. Ossa palatina	D. Astragal	IV. Formation of the Calyx
d. Ossa zygomatica	E. The Body of the Column above	V. Formation of the Corolla
e. Ossa lacrymalia	F. The Body of the Column below	VI. Formation of the Stamens
f. Ossa nasi	G. Listella, or Cincture, or Annulet	VII. Nectarines
g. Ossa frontis	H. Torus, or Tore	VIII. Further remarks of the Stamens
h. Os sphenoideum anterius	I. Orlo, or Plinth	IX. Formation of the Style
i. Os ethmoideum	J. Pedestal, or Stylobatum	X. Of the Fruits
j. Conchae		XI. Of the Coverings Lying Next to the Seed
etc.		etc.

All new or existing descriptions of species fall into an anatomical classification and nomenclature,[16] namely, a hierarchy of groups in which each named and defined character resides across all the developmental stages of the organism (i.e., metamorphosis). New species are compared against existing species and new manifestations of characters are described using a Typus, as opposed to a name-bearing type. In trilobites, for example, the Typus may be divided into,

1. The cephalon or head
   A. glabella
      a. glabellar furrows
      b. axial furrows
      c. preglabellar field
   B. Palpebral area
      a. palpebral lobes
2. The thorax or body
3. The pygidium or tail
   A. Pygidial axis
      a. axial rings
   B. Pygidial field
      a. pleural ribs
   C. Pygidial margins

The diagnosis for a species, *Lorrettina macrops* Shergold, 1972, would read,

> Glabella moderately tapered forward and strongly inflated. Three pairs of glabellar furrows. Axial furrows bowed slightly to moderately outward. Narrow (sag.), convex preglabellar field. Palpebral lobes long (exsag.), positioned medially between S2 and midpoint of L1; palpebral areas extremely narrow (tr.). Pygidium trapeziform. Pygidial axis short (sag.) with four axial rings. Pygidial pleural field with two pleural ribs defined by moderately impressed furrows. Pygidial posterior margin gently indented.
>
> *(Smith 2024, p. 55; Figure 5.2)*

Goethe's Typus is not a strange new method that has been lost in the mists of time. Rather it is a well-known, but undocumented, methodology used by taxonomists worldwide across most, if not all, taxonomic groups. Goethe's Typus is an abstract outline that acts as a checklist for taxonomists to "facilitate a comparative anatomy", without which "we will be unable to form a general picture of mammals".[17]

Goethe's Typus is very much alive and well in the 21st century, as are Anschauung and the Urphenomena, yet partitioners barely mention any *method* of comparative biology, or are ignorant of them entirely.

The Typus is an important part of the morphological method even though it is mostly an abstract tool. While it can be used to compare organisms of the same taxonomic class (e.g., mammals), it can also be used to compare organisms across other classes.

> If we wish to widen our osteological knowledge to all mammals, we can, through our method, approach other classes, such as the amphibians and birds [...].

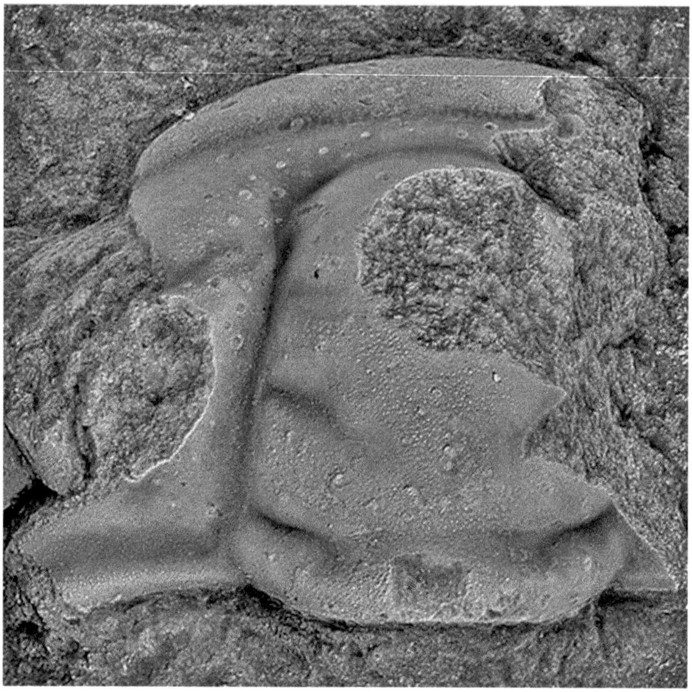

FIGURE 5.2 Image of the cephalon (head) of the dokimocephalid trilobite *Lorrettina waterhousei.*

(Photo by Patrick Smith used with permission.)

That osteological knowledge, however, pertains to all developmental stages of the organism.

> It is known that the human foetus and a newly born child contain more bone divisions than you find in an adolescent, fully grown or elderly person.[18]

The unfused bones in foetuses and babies show more of the osteology than in a human adult, and bones such as the intermaxillary are more likely to be found in humans if you were to examine the skull of a newborn. Goethe also mentions that the non-osteological parts, such as the muscles, nerves and vessels, may also be used to compare individual specimens across different classes.

Goethe stresses the differences in the number of parts (serialisation) when comparing different organisms as well as the different types of manifestations of parts.

> The greatest difficulty, however, arises from the fact that the bones of different animals are highly different in form. Therefore, the observer finds himself in a dilemma, whether he has entire skeletons in front of him or only individual parts. If he finds the parts out of context, he often doesn't know how to explain them; even if he has recognised them, he doesn't know how to describe them, and especially how to compare

> them [...] Who, for example, would consider the forearm of a mole and that of a hare to be the same part of related organisms? However, with regard to the way in which similar limbs of different animals can differ so greatly in form, and this will only become clear to us when we implement the method, we will first of all note the following in particular.

Goethe informs us that the same form can occur in different manifestations, such as the intermaxillary bone in deer and lions, which can be quite unlike each other.

> Thus, a bone can indeed be formed in a certain way, but through the remaining development, it can be compressed and misshapen to such an extent that one would hardly dare to recognise it as the same bone.[19]

Goethe admits that his osteological outline [Altas] would need various supplements in order to showcase all the manifestations that we can see as observers. The outline, including the places where these bones occur within an anatomy, is essential in identifying these manifestations of form, regardless of whether they have moved, or become compressed or formed in odd ways, thus allowing the comparative anatomist to "be able to derive and abstract the possible manifestations [Abweichungen] of this form partly from the concept and partly from experience".[20]

Goethe's Morphology is now complete; the method includes the activity of Anschauung and the Urphenomenon (including metamorphosis, polarity and Steigerung), and the abstract, namely the Typus, that together discover the whole. A less discerning reader of Goethe may jump to the conclusion that he had somehow "discovered" evolutionary theory (e.g., Wells 1978). I beg to differ. Goethe had attempted to describe the *method* of observing and comparing organisms and other natural objects, such as minerals,[21] which he called Morphology, rather than come up with model that *explains* the origins of organic form. At no time did Goethe propose or describe the processes responsible for biological form. For Goethe, phenomena and experience rise above abstract theories and mere speculations, even though it plagued his geology. For example,

> But when I now learned, that, to weaken the tradition of a [Biblical] deluge, he [Voltaire] had denied all petrified shells, and only admitted them as *lusus naturæ*, he entirely lost my confidence;[22] for my own eyes had, on the Baschberg [Bastberg near Bouxwiller, France], plainly enough shown me that I stood on the bottom of an old dried-up sea,[23] among the *exuvice* of its original inhabitants. These mountains had certainly been once covered by waves, whether before or during the deluge did not concern me: it was enough that the valley of the Rhine had been a monstrous lake, a bay extending beyond the reach of the eyesight; out of this I was not to be talked. I thought much more of advancing in the knowledge of lands and mountains, let what would be the result.[24]
>
> *(Goethe, trans. Oxenford, p. 420; also see LA I 1, p. 2)*

Assumptions, such as those proposed in his defence of Neptunism, never appeared in either his work on colour or in his botanical and zoological works.

METAMORPHOSIS OF ANIMALS

In 1796 to 1798 Goethe applied his learning-by-doing approach by observing the metamorphosis of several species of butterflies and moths; less so frogs, snails, mussels, beetles, bumblebees and worms. The approach involved gathering eggs or larvae and rearing them in captivity, observing their growth and dissecting them at each stage. The aim of Goethe's metamorphosis of animals was to observe the complete cycle of development from egg to adult, something that would have required far more material and time if applied to a vertebrate (see below).

The point of Goethe's metamorphosis of both animals and plants is to provide a description of the Urphenomenon, rather than a short diagnosis of the organism, which was mainly practised by Linnaean taxonomists.[25] Goethe's description of the development of the Sphingid Spurge hawk-moth (*Hyles euphorbiae*) is excruciating in its detail. Goethe lists 235 observed characteristics (and 22 follow-up notes), starting with fertilised and unfertilised eggs viewed under a microscope, right to end of the life cycle. The parts of each developmental stage were examined, including the head, thorax, legs, genitalia, wings, along with their various behavioural characteristics, such as moulting. Various measurements, such as the length and weight, were made and recorded.

The description of the Spurge hawk-moth (a Sphingid) differs from the *Metamorphose* in that it focuses on a single species rather than plants as a whole. The metamorphosis of animals does not leave out any anatomical features, such as the roots and tubers omitted in the *Metamorphose*. Goethe's metamorphosis of animals would have provided a far more complete picture of Sphingid development and behaviour than it did of plant development. Perhaps this is why Goethe's latter work on plant behaviour, such as the "Spiral tendency in vegetation", could be seen as addendums to his 1790 monograph.

Goethe's metamorphosis of animals was never published as a comprehensive study, rather it was compiled from several notes and bound in a bundle titled "The metamorphosis of insects, particularly butterflies, as well as their other characteristics and economy".[26]

Goethe makes few pronouncements in his outline of animal metamorphosis, however with one exception,

> [m]oulting is the most important phenomenon on which the metamorphosis of insects is based. All organisms work in hidden, covered and withdrawn beneath a cuticle, that can never be viewed as entirely simple, but rather made of many layers. This cuticle is not superfluous structure but a pure manifold vessel in which the organisation is contained. A vessel from which the content emanates.[27]

The vessel is analogous to an Urpflanze or Urtier, a vessel that spurts forth all the multifaceted forms of an organism. Essentially this could be said about a seed or an egg, but the pupae is an intriguing phenomenon, simply because what emerges or moults isn't anything like the initial form. A comparison could be drawn between the pupae and a tadpole, given the severity of the transformation of the juvenile form into the adult form (see Holdrege 2017). Goethe did observe and record the anatomy of frogs but only compared the male and female adult forms.

NOTES

1 *Dem Menschen wie den Tieren ist ein Zwischenknochen der obern Kinnlade zuzuschreiben* (LA I 9, pp. 154–161).

2 "Welch eine Kluft zwischen dem o*s intermaxillare* der Schildkröte und des Elefanten, und doch läßt sich eine Reihe Formen dazwischen stellen die beide verbindet" (LA I 9, p. 161; translated by Douglas Miller in Goethe 1995, pp. 115–116).

3 Thompson (1917) used mathematical laws to affect change in organisms that have been drawn over a grid. As the size of each of the grids change so does the form of the depicted organism.

4 Several more Ur-terms make there way in Goethe's work, including "Urform" (LA I 9, p. 288), "Urstier" ("fossil steer" LA I 9, p. 256), "Urworte" (LA I 9, p. 87) and "Urgeographie" (LA I, p. 242).

5 For a history of the usage of the term, see https://woerterbuchnetz.de/?sigle=DWB&lemid=U13668.

6 "Der Marmor ist ein seltsames Material, deswegen Apollo von Belvedere in Urbilde so grenzlos erfreulich, denn der höchste Hauch des lebendigen, junglingsfreien, ewig jungen Wesens verschwindet gleich in besten Gipsabguß" (Rome, 25 December 1786; WA I 30, p. 238).

7 "… es ist also auch damit wie mit Glück und Weisheit, davon uns die Urbilder nur vorschweben, deren Kleidsaum wir höchstens berühren" (Rome, 24 November 1787; WA I 32, p. 140).

8 Dies also hätten wir gewonnen, ungescheut behaupten zu dürfen. daß alle vollkommnern organischen Naturen, worunter wir Fische, Amphibien, Vögel, Säugetiere und an der Spitze der letzten den Menschen sehen, alle nach einem Urbilde geformt seien, das nur in seinen sehr beständigen Teilen mehr oder weniger hin und her weicht und sich noch täglich durch Fortpflanzung aus- und umbildet. (LA I 9, p. 198)

9

> Alle Glieder bilden sich aus nach ew'gen Gesetzen,
> Und die seltenste Form bewahrt im geheimen das Urbild.
>
> *(WA I 3, p. 89)*

10

> PROMETHEUS.
> Entwandelt friedlich! Friede findend geht ihr nicht.
> Denn solches Los dem Menschen wie den Tieren ward,
> Nach deren Urbild ich mir Beßres bildete …
>
> *(WA I 50, p. 310)*

11 Hiebei fühlte ich bald die Notwendigkeit einen Typus aufzustellen, an welchem alle Säugetiere nach Übereinstimmung und Verschiedenheit zu prüfen wären, und wie ich früher die Urpflanze aufgesucht, so trachtete ich nunmehr, das Urtier zu finden, das heißt denn doch zu letzt, den Begriff, die Idee des Tieres. (LA I 9, p. 13)

12 Douglas Miller uses the term "archetype". Given the loaded nature of this term (i.e., ancestor, prototype, bauplan, etc.), I will use Goethe's original term instead.

13 "Die Erfahrung muß uns vorerst die Teile lehren, die allen Tieren gemein sind, und worin diese Teile verscheiden sind" (LA I 9, p, 121; Translated by Douglas Miller in Goethe 1995, p. 118).

14 Deshalb geschieht hier Vorschlag zu einem anatomischen Typus, zu einem allgemeinen Bilde, worin die Gestalten sämtlicher Tiere, der Möglichkeit nach, enthalten wären, und wonach man jedes Tier in einer gewissen Ordnung beschriebe. Dieser Typus müßte so viel

wie möglich in physiologischer Rücksicht aufgestellt sein. Schon aus der allgemeinen Idee eines Typus folgt, daß kein einzelnes Tier als ein solcher Vergleichungskanon aufgestellt werden könne: kein Einzelnes kann Muster des Ganzen sein. (LA I 9, p. 121; Translated by Douglas Miller in Goethe 1995, p. 118)

15 Pierre Belon (1517–1564) offered a similar method in which all the skeletal parts of a human and a bird were labelled accordingly. Belon (1555), however, did not group these characters into anatomical units (e.g., the head or rear section) or propose a Typus from which both organisms can be compared.

16 I use the term nomenclature to refer to the names taxonomists give to body parts, rather than to the naming systems of taxa as used, for example, in the International Code of Zoological Nomenclature.

17 "... Erleichterung der verglichenden Anatomie", "... weil wir ohne eine solchen Überblick kein allgemeines Bild der Säugetiere aufstellen können ..." (LA I 9, p. 120, 122).

18 Wollen wir nun wir nun aber unsere osteologischen Kenntnisse über dis sämtlichen Säugetiere ausbreiten, wollen wir dabei so zu Werke gehen daß wir durch unsere Methode selbst den anderen Tierklassen, den Amphibien und Vögeln, uns nähern [...] Es ist bekannt daß schon beim menschlichen Fötus und bei einem neugebornen Kinde sich mehrere Knochenabteilungen finden als bei einem Halberwachsenen, und bei diesem wieder mehr als bei einem ausgewachsenen oder veralteten Menschen. (LA I 9, p. 137)

19 Es tritt nun aber die größte Schwierigkeit ein, welche daher entspringt, daß auch die Knochen verschiedener Tiere einander in der Form höchst unähnlich sind. Daher gerät der Beobachter, mag er ganze Skelette vor sich haben oder nur einzelne Teile, gar oft in Verlegenheit. Findet er die Teile außer dem Zusammenhange, so weiß er oft nicht wofür er sie erklären soll; hat er sie aber auch erkannt, so weiß er nicht wie er sie beschreiben, und insonderheit wie er sie vergleichen kann [...] Wer würde z.B. den Oberarm eines Maulwurfs und des Hasens für eben denselben Teil verwandter organiser Wesen halten? Von den Arten jedoch wie gleiche Glieder verschiedener Tiere in der Form so sehr von einander abweichen können, und die uns erst bei der Ausführung ganz deutlich werden dürften, wollen wir uns vorerst folgende vorzüglich merken [...] So kann ein Knochen zwar in einem gewissen Sinne ausgebildet, aber durch die übrige Bildung zusammengedrängt und mißgestaltet sein, daß man gleichfalls kaum wagen würde ihn für denselbigen Knochen zuerkennen. (LA I 9, pp. 140–141)

20 "Man wird alsdann die möglichen Abweichungen dieser Form teils aus dem Begriff, teils aus der Erfahrung herleiten und abstraheiren können" (LA I 9, p. 142).

21 Goethe understood that organisms had Mineral-Körper or mineral bodies, such as those that form teeth and bones.

22 It is likely that Goethe is referring to Voltaire's *La dissertation sur les changements arrivés dans le globe* (Voltaire 1746), rather than *Les Singularites de la Nature* (Voltaire 1768). Goethe ascribes Voltaire's dismissal of scriptural geology, namely the use of scientific methods to prove the Biblical flood, as rejecting the notion of fossils and palaeoenvironments entirely. It was harsh for Goethe to castigate the French philosopher, given both were advocating the same thing, namely for naturalists to be careful to describe what they observe rather than to simply assume (Carozzi 1983). For example, Voltaire claimed,

> I only desire to be permitted to examine, according to the rules of probability, whether this globe either has been, or will one day be, entirely different from what it now is. And here we have nothing more to do than make use of our eyesight.
>
> *(Voltaire 1746 [1901], p. 226)*

23 Goethe's "dried up old sea" was in fact an Eocene freshwater lake system (lacustrine basin; see Schumacher 2002).

24 Da ich nun aber gar vernahm, daß er, um die Überlieferung einer Sündflut [Sintflut] zu entkräften, alle versteinten Muscheln Leugnete und solche nur für Naturspiele gelten ließ, so verlor er gänzlich mein Vertrauen: Denn der Augenschein hatte mir auf dem Baschberge deutlich genug gezeigt, daß ich mich auf altem abgetrockneten Meeresgrund, unter den Exuvien seiner Ureinwohner befinde. Ja! diese Berge waren einstmals von Wellen bedeckt; ob vor der während de Sündflut [Sintflut], das konnte mich nicht rühren, genug, das Rheintal war ein ungeheurer See, eine unübersehliche Bucht gewesen; das konnte man mir nicht ausreden. Ich gedachte vielmehr in Kenntnis der Länder und Gebirge vorzuschreiten, es möchte sich daraus ergeben was da wollte. (LA I 1, p. 2)

25 The comparison may seem unfair, but it highlights the difference between naming and identifying of an organism versus observing and describing the development of an individual. The latter led to comparative biology, while the former is crucial to taxonomy.

26 "Die Metamorphose der Insekten, besonders der Schmetterlinge, wie auch ihre übrige Eigenschaften und Ökonomie betreffend" (LA II 9B, p. 447).

27

> Das Häuten ist das wichtige Phänomen, vorauf die Metamorphose der Insekten beruht. Alles Lebendige wirkt in Verborgenen, bedecket, verschlossen, unter einer Haut, die niemals als ganz einfach angesehen werden kann, sondern aus mehren Lamellen besteht. Diese Haut aber ist nicht etwa ein drittes, überflüssiges, sondern das reine, mannigfaltige Gefäß, worin die Organisation enthalten ist. Ein Gefäß aus dem Inhalt entsprungen.
>
> *(LA I 10, p. 171)*

REFERENCES

Belon P (1555) *L'histoire de la Nature des Oyseaux*. Guillaume Cavellat, Paris.

Carozzi M (1983) Voltaire's Attitude Toward Geology. *Archives des sciences et compte rendu des séances de la Société*, 36: 1–145.

Ebach MC (2022) *Goethe in the Age of Artificial Intelligence*. Palgrave Macmillan, Singapore.

Goethe JW (1995) Scientific studies (Vol. 12). *Goethe: The collected works* (Ed. D Miller). Princeton, New Jersey.

Holdrege C (2017). *Do frogs come from tadpoles?* Nature Institute, Ghent, New York.

Palladio A (1742). The architecture of A. Palladio, in four books: Containing a short treatise of the five orders, and the most necessary observations concerning all sorts of building: As also the different construction of private and public houses, high-ways, bridges, market-plates, xystes, and temples, with their plans, sections, and uprights. A. Ward.

Schumacher ME (2002) Upper Rhine Graben: Role of preexisting structures during rift evolution. *Tectonics*, 21: 1–17. https://doi.org/10.1029/2001TC900022

Shergold JH (1972) Late Upper Cambrian trilobites from the Gola Beds, western Queensland. Bureau of Mineral Resources. *Geology and Geophysics, Australia, Bulletin*, 112: 1–89.

Smith PM (2024) A new species of *Lorrettina* (dokimocephalid trilobite) from the Iverian (Cambrian, Jiangshanian) of the Amadeus Basin, Northern Territory, Australia. *Alcheringa*, 48: 52–60.

Stieler K (1691) *Der Teutschen Sprache Stammbaum und Fortwachs*. Johann Hofmanns, Nürnberg.

Thompson DW (1917) *On Growth and Form*. Cambridge University Press, Cambridge.

Voltaire F-MA (1746) The changes in our globe [La dissertation sur les changements arrivés dans le globe]. *Voltaire: Essays on Literature, Philosophy, Art History*, 19: 279–292.

Voltaire F-MA (1768) *Les Singularites de la Nature*. Chez Huzard, Paris.

Wells GA (1978) *Goethe and the development of science 1750–1900*. Sijthoff and Noordhoff, Alphen can den Reign, The Netherlands.

6 Reimagining Goethe's Art of Science

Goethe's science has been claimed by Ernst Haeckel for evolutionism and Anthroposophists for Goetheanism. Goethe was not an evolutionist and neither was he, rather ironically, a Goetheanian. Some researchers – Jakob Uexküll (1864–1944), Wilhelm Troll (1897–1978) and Adolf Portmann (1897–1982) – recognised the atheistic and comparative nature of Goethe's Morphology. Goethe's work, viewed by a practising scientist, is recognisable as comparative biology. While his work is known from fragmentary notes or in collections with no formal structure, most taxonomists will recognise it for what it is, namely, how organisms are compared and described *aesthetically*. Goethe's work was a prototype for what would become a large and sprawling scientific field.

Towards the end of his life, natural history was morphing into a *science* and becoming more specialised and professional. It was only a year after his death in 1832 that the term science was first used by William Whewell (1794–1866). The first scientific journals specialising in fields of zoology, such as entomology, were founded around the same time. The scientific method was also becoming formalised and methodologies developed within specialties, such as taxonomic nomenclature. In Goethe's day professors in medicine would study a variety of plants and animals and teach general courses on natural history. They attended small gatherings to deliver lectures on nature that they would later publish in literary magazines such as the *Teutsche Merkur*, alongside poems, short stories and taxonomic treatments. Not long after, academia became more like that of the 20th century, where a professor of zoology would teach embryology to PhD students, attend scientific conferences and publish in dedicated scientific journals.

The professionalisation of science resulted in increasing specialisation, which scorned natural history as amateurish or generalist. The polymath was replaced by the *Fachidiot*, an expert with little knowledge beyond their own specialty. That trend continues today. Goethe the polymath, who was a scientist, poet, novelist, screenwriter, theatre director, minister of state and fulfilled many other administrative roles, was impossible to place in the era of professionalism, and over time he became best known only as a poet, and his scientific work was largely forgotten. If scorn was heaped onto natural historians by scientists, what would they think of a poet doing the same? British neurophysiologist Charles Sherrington (1857–1952) tells us,

> Were it not for Goethe's poetry, surely it is true to say we should no longer trouble about his science [...] Goethe, though devoted to science, had not at root the scientific temperament. He had not for instance, along with the urge to discovery the sublime detachment of the scientific thinkers.
>
> *(Sherrington 1949, pp. 29–30)*

DOI: 10.1201/9781003275343-6

One wonders who these scientific thinkers were? Sherrington was writing in the 1940s and by then Leonardo da Vinci had been called the "Father of Palaeontology" (Clements 1949, p. 1937) and Aristotle the "father of biology" (Ritter 1932, p. 377).

Reading Goethe, it is important to understand what he practised as opposed to what he thought. Medawar was careful to warn future historians of science that,

> What scientists do has never been the subject of a scientific, that is, ethological inquiry […]. It is no use looking to scientific 'papers', for they not merely conceal but actively misrepresent the reasoning that goes into the work they describe […].
>
> *(Medawar 1968, p. 15)*

The same can be said for Goethe's writings (Table 6.1). In his geological works we can see him struggling to fit phenomena he observes into Neptunian theory that is unable to explain many observed geological formations. Much of Goethe's musings are in drafts, recorded from conversations, letters or notes. We can't expect Goethe to have formalised his ideas until they're published, and even then we need to be careful about how often he changes his mind or uses different terms for the same concept. To best appreciate Goethe's scientific works is to remove the thinking and pay attention to what he practised, the terms he used and didn't use, as well as the mistakes he made. Most of all Goethe needs to be recast as a mainstream scientist who influenced and contributed to comparative biology and colour science.

TABLE 6.1
A sample of published works of Goethe including the year of composition to show that many of his methodological studies in colour, zoology and botany were written at the same time

Year of Composition	Title of Work and Reference	Year Published
1781–1782	Nature [Die Natur, Fragment LA I 11, pp. 3–5]	1782–1783
1784	On Granite [Über den Granit LA I 1 pp. 57–63]	1877
1784–1785	A Study Based on Spinoza [Studie nach Spinoza LA I 11 pp. 6–8]	1891
1784–1786	*An Intermaxillary Bone is Present in the Upper Jaw of Man as Well as in Animals* [Dem Menschen win den Tieren ist ein Zwischenknochen der obern Kinnlade zuzuscheriben LA I 9, pp. 154–161]	1820
1789	Suggestions for a Comparative Approach Reconciling the Plutonists and Neptunists on the Question of Basalt [Vergleichs Vorschlage die Vulkanier und Neptunier über die Entstehung des Basalts zu vereinigen LA I 11, pp. 37–38]	1892
1790	*The Metamorphosis of Plants [Die Metamorphose der Pflanzen* LA I 9, pp. 23–62]	1790
1790–1794	Toward a General Comparative Theory [Versuch einer allgemeinen Vergleichungslehre LA I 10 pp. 118–122]	1892
1791	*Contributions to Optics I [Beiträge zur Optik I* LA I 3]	1791

(Continued)

TABLE 6.1 (Continued)

Year of Composition	Title of Work and Reference	Year Published
1792	*Contributions to Optics II [Beiträge zur Optik II* LA I 3]	1792
	The Experiment as Mediator between Object and Subject [Der Versuch als Vermiittler von Objekt und Subjekt LA I 3, pp. 285–295 & LA I 8, pp. 305–315]	1823
1794	Fortunate Encounter [Glückliches Ereignis LA I 9, pp. 79–83]	1817
	The Extent to which the Idea "Beauty is Perfection in Combination with Freedom" may be applied to Living Organisms [I wiefern die Idee: Schönheit sei Vollkommenheit mit Freiheit, auf organische Naturen angewendet werden könne LA I 10, pp. 125–127]	1953
	Tibia and Fibula [Tibia and Fibula LA I 9, pp. 365–368]	1824
1795	*Outline for a general Introduction to Comparative Anatomy, Commencing with Osteology* [Erster Entwurf einer allgemeinen Einleitung in die vergleichende Anatomie, ausgehend von der Osteologie LA I 9, pp. 119–151]	1820
	Observation on Morphology in General [Betrachtung über Morphologie überhaupt LA I 10, pp. 140–144]	1891
	Studies for a Physiology of Plants [Versuche zur Methode der Botanik: Ordnung des unternehmens LA I 10, pp. 129–131]	1891
1798	*Empirical Observation and Science* [Das Reine Phänomen LA I 11, pp. 39–40]	1893
1805	Lectures on Physics: Introduction [Physikalische Vorträge schematisiert: Einleitung LA I 3, pp. 416–417, LA I 11, p. 55–56]. Also known as Polarity [Polarität WA II 11, pp. 164–166]	1893
	Symbolism [Symbolik LA I 11, pp. 56–57, version in LA I 3, pp. 417–418 is incompete]	1893
1807	The Enterprise Justified [Das Unternehmen wird entschuldigt LA I 9, pp. 5–6]	1817
	The Purpose Set Forth [Die Absicht eingeleitet LA I 9, pp. 6–10]	1817
1808	*Farbenlehre*: Didactic Part [Didaktischer Teil LA I 4]	1808
1808	*Farbenlehre*: Polemic Part [Polemischer Teil LA I 5]	1810
1809	*Farbenlehre*: Historical Part [Historischer Teil LA I 6]	1810
1809	*Farbenlehre*: Confessions of the Author [Konfession der Verfassers LA I 6 pp. 412–429]	1810
1810	Announcement and Survey [Anzeige und Übersicht des Goethischen Werkes zur Farbenlehre LA I 7, pp. 1–17]	1810
1810	Theory of Tone [Tonlehre LA I 11, pp. 134–138]	1834
1813	Double Images of Calcite Crystals [Doppelbilder des rhombischen Kalkspats LA I 8, pp. 16–20]	1817
1816–1817	The Content Prefaced [Der Inhalt bevorwortet LA I 9, pp. 11–14]	1817
1817	Judgement through Intuitive Perception [Anschauende Urteilskraft LA I 9, pp. 95–96]	1820
	The Influence of Modern Philosophy [Einwerkung der neueren Philosophie LA I 9, pp. 90–94]	1820
	Elements of Entopic Colours [Elemente der entopischen Farben LA I 8, pp. 21–24]	1817

(Continued)

TABLE 6.1 (Continued)

Year of Composition	Title of Work and Reference	Year Published
1817–1818	The Formative Impulse [Bildungstrieb LA I 9, pp. 99–100]	1820
1817–1820	Colours in the Sky [Farben des Himmels LA I 11, pp. 199–200]	1879
1818	Doubt and Resignation [Bedencken und Ergebung LA I 9, pp. 97–98]	1820
1820	A Friendly Greeting [Freundlicher Zuruf LA I 9, p. 222]	1820
	My Relationship to Science, and to Geology in Particular [Verhältnis zur Wissenschaft, besonders zur Geologie LA I 11 pp. 215–217]	1873
	Entopic Colours [Entopische Farben LA I 8, pp. 94–138]	1820
	Chromaticism (Addenda) [Chromatik (Nachträge) LA I 8, pp. 342–343]	1822
1820–1821	Perception in the Subjective View of Purkinje 1819 [Das Sehen in subjektiver Hinsicht von Purkinje 1819 LA I 9, pp. 343–352]	1824
1821	The Eye: Receptive and Reactive [Auge/empfänglich und gegenwirkend LA I 8, p. 177]	1822
1822	Johann Kunckel The Art of Glassmaking [Johann Kunckel LA I 8, pp. 316–319]	1823
	Luke Howard to Goethe: a Biogeographical Sketch [LA I 9, pp. 264–265]	1822
	Remarkable Healing of a severely Damaged Tree [Merkwürdige Heilung eines schwer verletzten Baumes LA I 9, p. 238]	1822
1823	New Entopic Case (Ice Flowers) [Neuer endoptischer Fall (Eisblumen) LA I 8, p. 344]	1823
	Problems [Probleme LA I 9, pp. 295–297]	
	Significant Help Given to by an Ingenious Turn of Phrase [Bedeutende Fördernis durch ein einziges geistreiches wort LA I 9, pp 307–310]	1823
	A General Observation [Allgemeine Betrachtung LA I 9, p. 317]	1823
1824	Ernst Stiedenroth: A Psychology in Clarification of Phenomena from the Soul [Ernst Stiedenroth: Psychologie our Erklärung der Seelenerscheinungen LA I 10, pp. 226–227]	1824
	An Unreasonable Demand [Unbillige Forderung LA I 10, p. 215]	1891
1825	Toward a Theory of Weather [Versuch einer Witterungslehre LA I 11, pp. 244–268]	1833
1826–1827	Natural Philosophy [Naturphilosophie LA I 11, pp. 284–285]	1827
1828	*Bignonia radicans* [*Bignonia radicans* LA I 10, pp. 253–255]	1891
1829	Analysis and Synthesis [Analyse und Synthese LA I 11, pp. 301–303]	1833
1831	The Spiral Tendency in Vegetation [Allgemeine Spiraltendenz der Vegetation LA I 10, pp. 339–342]	1834

Note that books are written in italics. German translation given in square brackets where necessary. Dates and English titles sourced from Matthaei in Goethe (1971, pp. 205–206) and Miller in Goethe (1995, pp. 313–314). *Zur Naturwissenschaft Überhaupt* was serialised between 1817 and 1824 as two volumes (1817–1822 and 1822–1824). References given in square brackets.

BRINGING GOETHE IN FROM THE COLD

Goethe's scientific writings have seemingly been sidelined because they have been misunderstood, mostly by people who do not work in comparative science, whereas biologists and colour theorists from the late 18th to the early 21st centuries have hailed Goethe as a visionary.

> I am sure most of my readers honour Goethe only as a poet and a man; only a few have any conception of the high value of his scientific works, and of the gigantic stride with which he advanced before his own age – advanced so much that most naturalists of that time were unable to follow him.
>
> *(Haeckel 1876, p. 81)*

> Goethe's scientific work, which laid the foundation of comparative anatomy and plant and animal morphology, continues its fruitful influence to the present day.
>
> *(Magnus 1949, p. 124)*

> Among other things, [Goethe] was an accomplished botanist, he helped found the field of comparative anatomy, he coined the term *morphology*, and he anticipated the theory of evolution. If these achievements do not sound enough like "hard science," it is well to remember that in 1784 Goethe discovered the intermaxillary bone in the human jaw, thus supplying a link to primate anatomy that proved crucial to later evolutionary theories.
>
> *(Cantor 2004, p. 75)*

If Goethe has had such a huge impact on comparative biology and colour science, why is his science labelled as holistic, spiritual and Romantic? These terms indicate that Goethe's science is unique, not part of mainstream science; it's as though Goethe and his Morphology have been left out in the cold.

In order to bring Goethe's scientific writing back in from the cold we need to understand that his method, Morphology, is still practised in comparative biology today. We also need avoid labels that hinder, rather than celebrate, his work, as well as recognise where Goethe's science is wrong or misleading.

Importantly, we need to keep in mind that Goethe was practising outside of an academic institution in a period when science was becoming formalised and professionalised. Humboldt seemed to have survived this professionalisation, possibly because his followers, the Humboldtians, had developed the science of ecology seemingly independently of taxic regionalisation (see Ebach 2015). Goethe, unlike Humboldt, didn't publish most of his works in French, or in widely read journals or book series, nor did he attend or present at scientific conferences. Goethe wasn't as widely read due to his work being inaccessible to non-German readers; in fact much of his scientific work is still untranslated.

GOETHE'S MORPHOLOGY RETOLD

Goethe described how taxonomists and comparative anatomists practised their science. He called this Morphology and understood that it was a method that can be applied across all comparative science, be it anatomy, mineralogy or colour theory. Given that observation of phenomena involves a high degree of aesthetic appraisal (*sensu* Steigerwald 2002), Goethe's Morphology can be said to encompass both art and science, hence the title of this book *Goethe's Art of Science*.

Aesthetic appraisal, and our ability to use it as a scientific method, is often dismissed as "subjective" and more to do with folk classification than with real science. Yet, when we look at any taxonomic description or revision, we see aesthetic appraisal. Take the diagnosis of the harpetid trilobite *Eskoharpes palanasus* for example,

> Cephalon sagittally relatively short and wide, maximum width opposite middle of genal area. Brim flat, moderately wide and gently declined. External rim broad, overhanging internally. Marginal band concave. Genal roll moderately steep. Prolongations relatively short and convergent distally.
>
> *(McNamara et al. 2009, p. 15)*

None of these characters have been measured "objectively" or against a standard. What is the standard for "gently declined" brim or short versus long prolongations? There doesn't need to be one! We can see that the brim of *E. palanasus* is gently declined and the prolongations are short, when compared with other species of harpetids. Comparison involves a high degree of aesthetic appreciation. When we see a steep genal roll we know, as palaeontologists, that it is generally unusual in harpetids and we are enthralled. When we find new characteristics it excites us because we see the beauty in these organisms. I hear taxonomists talk about how "wonderful" or "beautiful" their organisms are, even those working on cockroaches! Without any aesthetic appraisal we wouldn't go near a cockroach, spider or naked mole rat. While this form of aestheticism seems unscientific, it is one of the main reasons why we work on the organisms that we do. In other words, it would seem odd that an arachnophobe would observe, touch or dissect a spider under a microscope, let alone breed them in tanks as do arachnologists.

Denying that taxonomists and comparative biology practise aesthetic appreciation is as absurd as saying that experts on Renaissance paintings don't find them beautiful. The division between science and art has left aesthetic appreciation as something that only happens in the arts. While it is fine for an archaeologist to find Greek vases beautiful and identify them based on scenes depicted on them, we may insist, as Sherrington does, that scientists shun such "subjective" acts and get on with the "sublime detachment of scientific thinkers". Surely there is some degree of aesthetic appreciation in the work that scientists do? What of the passion and commitment to science communication, be it an interview on the radio or a keynote lecture at a conference? A dispassionate talk read by a detached scientist would kill the subject matter. What of the students who wish to continue onto a PhD? What drives them? Many scientists I meet are passionate about their science and do describe natural phenomena in aesthetic terms, such as "wonderful" and "beautiful". Rather than Sherrington's dispassionate drones, scientists simply don't mention their aesthetic appraisal of the phenomena on which they work. The same is true with all observation in science: we fail to mention the person or people doing the observation. We may legitimately ask how they are observing, how do they recall previous phenomena and so on.

Anschauung is perhaps the most disciplined act of observation possible. A trained taxonomist can see trilobites in longitudinal cross-section in a lump of limestone rock. They look like a shepherd's crook. Once you see a few, you can also get your eye in and find them yourself. Training someone for just a few moments means that these little squiggles now are meaningful cross-sections. After four years of training, the time it takes to do a PhD in trilobite taxonomy and anatomy, you can be skilful enough to recognise trilobite fossils in any observable stratigraphic sequence. Even in specimens that have been almost ground down, you are able to pick out the anatomy, in exactly the same way a palaeoanthropologist can pick up a flint flake and recognise that it was deliberately made by someone in the Palaeolithic period.

Anschauung is a vital part of training a taxonomist, archaeologist or mineralogist. The ability to understand what you are seeing makes these descriptive sciences so important because they give meaning to natural objects. Through Anschauung we can discover and classify the natural world because we are observing well beyond glancing at something. Anschauung is not restricted to sight. We have other senses that can be used to describe and classify the natural world, such as hearing, touch and smell. All these senses combined, if trained well, can help in understanding the natural world.

Concurrently with training someone to use their senses, they can be trained to recall accurately. The ability to recall in your mind's eye the many phenomena you have observed means that you can start to recognise them again, but as different manifestations. We might have only ever seen cats, but when we see kittens, we appreciate that what we are seeing is very much like a cat, but the characteristics are all altered in some way. We recognise the different parts of the cat's anatomy, but we see that the legs are less muscular or the head is wider and less elongate. If we look at a cat foetus we see the same structures again, but this time they are modified. If we see a cat embryo, we see the same structures but altered again. We retain these memories as living images that can be recalled at any time. If we have seen enough of these features, we can almost play the images in our mind and see the cat develop in our mind's eye. All those experiences that we have captured in our mind form the Urphenomenon. Imagine, if you will, observing the Moon every night for a month. That one cycle, namely, new Moon, waxing crescent, first quarter, waxing gibbous, full Moon, waning gibbous, third quarter and waning crescent, can be captured in our mind's eye as a type of Urphenomenon. If we have raised a pet from its birth to its death, we too have an Urphenomenon of that pet in our mind's eye. The Urphenomenon is a powerful tool, particularly in taxonomy and other comparative sciences, in identifying natural phenomena and in the discovery of new phenomena. If you specialise in harpetid trilobites, you will know all 250 or so species. If you observe one that you have never seen before, it might be a new species. There is no quantifiable way of doing this at the level of a trained observer, but an expert casting an eye over a freshly exposed slab of rock can identify something new quite quickly.

Anschauung and the Urphenomena are crucial to taxonomy, comparative biology, mineralogy and to many other historical sciences. Without developing the skills and knowledge (i.e., Anschauung and the Urphenomenon), we will find it hard to identify natural phenomena and discover new ones. Goethe understood this and sought to outline this as a method, namely Morphology, for comparative osteology, botany and *Farbenlehre*. The way Goethe developed and practised his Morphology was by proposing an idea, practising Anschauung and developing the Urphenomenon, the final goal of Morphology.

An early example of this approach was the Urpflanze (Chapter 4). Goethe first mentioned the Urpflanze in 1786 during his travels in Italy. The idea was simple: was it possible to have every possible manifestation of plants in one image? To develop his idea, Goethe needed to practise Anschauung by observing the development of plants, which led to his *Metamorphosis of Plants* in 1790. His learning-by-doing approach was ideal as each stage of a plant's development was recorded in detail, from the germinating seed to the fully grown plant and fruit. The same is true for his Urtier (Chapter 5), which was developed further through observing each life stage of the Spurge hawk-moth. In his *Beiträge*, Goethe's idea was that light came from darkness, a principle of his polarity. The learning-by-doing approach of using a tool, namely the glass prism, meant that he could see refraction of light along the white and black borders of his cards. In his comparative osteology, Goethe proposed the Typus to help establish an anatomical outline of a mammal from which he could compare all other mammals. The Typus was simply a tool that, like the prism, helped him achieve a systematic way to compare organisms.

The result of each of these ideas is that through Anschauung and the use of tools, we have developed the Urphenomenon in our mind's eye. Goethe's approach was the same in his *Farbenlehre*, botanical and zoological work, namely,

$$\text{Hypothesis}(\text{Wholeness}) -> \text{Anschauung}[+\text{Tools}] -> \text{Urphenomenon}$$

The abstract idea, namely that of a Urpflanze, Urtier or polarity, was a hypothesis. It derives from general observation and logical argumentation. Anschauung and comparing metamorphosis in plants and animals, refraction through a prism or comparing bones via a Typus was the method. The Urphenomenon was the result, namely,

$$\text{Hypothesis} -> \text{Method}[+\text{Tools}] -> \text{Phenomenal Result}[+\text{Physical Result}]$$

In each case we have:

Colour

$$\text{Polarity} -> \text{Colour / Turbidity}[+\text{Prism}] -> \text{Urphenomenon}[+\text{Description}]$$

Botany

$$\text{Urpflanze} -> \text{Metamorphosis}[+\text{Tools}] -> \text{Urphenomenon}[+\text{Description}]$$

Zoology A

$$\text{Urtier}->\text{Comparative Osteology}\left[+\text{Typus}\right]->\text{Urphenomenon}\left[+\text{Description}\right]$$

Zoology B

$$\text{Urtier}->\text{Metamorphosis}\left[+\text{Tools}\right]->\text{Urphenomenon}\left[+\text{Description}\right]^{1}$$

In McNamara et al. (2009) (quoted above), we have the aim, namely "Is this a new species?", and the physical result, a diagnosis and description. What is missing is the method, namely Anschauung, and the phenomenal result – the Urphenomenon. Why are these missing? For the most part taxonomists don't consciously realise they are practising it, in the same way most scientists don't realise that they practise aesthetic appraisal. Or, if aware of it, they don't think it is important or it is too subjective and borders on the so-called "mystical".[2] In most part, I think practitioners of Anschauung and the Urphenomenon are simply unaware that these are just as important as the physical description. Goethe's wish that this become front and centre of Morphology was lost on many of his contemporaries. Science has become obsessed with objectivity, to the point that we are far happier to implement and trust the results of computer models that nobody understands than trust our own abilities to observe and recall. In this sense Morphology is composed of the abstract idea, namely Wholeness; what we practise, namely Anschauung; and the recall of our experiences, that is the Urphenomenon:

$$\text{Morphology} = \text{Wholeness}\left(\text{Hypothesis}\right) + \text{Anschauung} + \text{Urphenomenon}\left(\text{Practise}\right)$$

Scientific method must reflect meaning, that is, understanding what you do as a scientist and understanding your experience of phenomena, be it organisms, minerals, rocks, clouds, colour or sound. If we understand our method and our phenomena then we find meaning within them. When we see someone recognise and identify a bird or a Picasso and hail that person as an expert, we miss the point. Understanding something means we are able to identify it and learn from it again and again. We observe it via Anschauung and it becomes part of us through the Urphenomenon. We understand that the phenomenon is a whole and that there is so much more to it than what we see in a given moment. We constantly learn from the phenomenon the more we observe it over time. Practising Anschauung, developing the Urphenomenon and understanding the whole is Goethe's Art of Science.

HOLISM, SPIRITUALISM AND ROMANTICISM

"Holism", "spiritualism" and "Romanticism" are terms that have been associated in the recent literature with Goethe's scientific writings. These terms were never used by Goethe to describe his work. "Holism" wasn't coined until the early 20th century, "spiritualism" is a mistranslation of the word "Giestens" and Goethe's scientific writing predates the concept of Romanticism, which was popularised in the 19th century.

The term holism was introduced by Jan Christian Smuts (1870–1950) in his book *Holism and Evolution* (Smuts 1927):

> Wholes are not mere artificial constructions of thought; they actually exist; they point to something real in the universe [...] Taking a plant or an animal as a type of a whole, we notice the fundamental holistic characters as a unity of parts which is so close and intense as to be more than the sum of its parts [...] the whole is in the parts and the parts are in the whole, and this synthesis of whole and parts is reflected in the holistic character of the functions of the parts as well as of the whole.
>
> *(Smuts 1927, p. 88)*

What Smuts was referring to is the Urphenomenon, which is both real and something we can see with our mind's eye (i.e., recall). Regardless, the whole is perceived by the observer as experiences, and these become the Urphenomenon (see above). The term holism has been adopted in different fields by different researchers and is used in different ways. For example,

> the belief that each thing is a whole that is more important than the parts that make it up.
>
> *(Cambridge Advanced Learner's Dictionary & Thesaurus)*

> a study or method of treatment that is concerned with wholes or with complete systems: a holistic study or method of treatment.
>
> *(Merriam-Webster.com Dictionary n.d.)*

Holism is used here as a synonym of "integrative" in the sense that we should concentrate on aspects of wholeness rather than just focus on a particular set of the parts. Integrative taxonomy, for example, proposes that all aspects of the whole should be used in taxonomic descriptions, including morphology, genomics, biochemistry, histology, biogeography and ecology. In this sense the whole is the sum of its parts, but you are using as many as possible to reach a complete whole. Another way of interpreting the "whole" is by claiming to have a data set in which n = all.

Holism has other definitions too, such as "the theory that parts of a whole are intimately interconnected, such that they cannot exist independently of the whole".[3] In this sense, we need to look at all aspects of health care, including the physiology and psychology of patients. Neither of these definitions refers to Urphenomenon; rather, they look at static systems and how their parts are interconnected.

Goethe's scientific work is often described as "holistic", yet Goethe never used the term "Holismus". Rather, he used other terms, such as "Einheit" or "Ganze" or "Ganzen", but these refer to wholeness via perception rather than holism. Goetheanians, however, use Goethe's definition of wholeness when they refer to "holism" (Holdrege 2005). Given that they are a minority when compared to the Wellness/Holism juggernaut for example, it seems that we are better off using Goethe's original term "Whole" instead.

"Spiritualism" is a term introduced, quite possibly by Rudolf Steiner, to describe Goethe's scientific method. The term is inferred from "Geist" or "geistig", which is often translated as "spirit" or "spiritual", which invokes a "*vitalis spiritus*", the life force of vitalism. In German this is termed "Lebenskraft". When Goethe uses the terms "Geist" [mind] or "geistig" [mental or intellectual] he is referring to the mind.

> Why should it not also hold true in the intellectual area that through Anschauung of eternally creative nature we may become worthy of participating mentally [geistigen] in its creative processes? [...] A living thing cannot be measured by something external to itself; if it must be measured, it must provide its own gauge. This gauge, however, is highly intellectual [geistig], and cannot be found through the senses ....[4]

The terms "spirit" and "spiritual" do not make much sense when we understand that Goethe is talking about the *mind*'s eye.

Another term that Goethe did not use to describe his scientific method is Romanticism. Goethe was quite explicit as to what he thought of Romanticism.

> The classical I call healthy, and the romantic sick.[5]
>
> *(Goethe to Eckermann, 2 April 1829)*

Compare this to another conservation a year later,

> [Schiller] proved to me that I myself, against my will, was romantic, my *Iphigenia*, through the predominance of feeling, am by no means as classical and in the ancient sense as one might perhaps believe. ...[6]
>
> *(Goethe to Eckermann, 21 March 1830)*

Perhaps Schiller was referring to Goethe's 1781 play *Iphigenia* and not to his scientific writing. Richards decided to use the latter remark as the leitmotiv for an entire book. Richards even admits that most historians of 19th-century science "usually dismiss anything sounding like Romantic science as an aberration" (Richards 2002, p. 3). If this is the case, why taint Goethe with the term "Romantic", a concept that is attributed to Friedrich Schlegel (1772–1829)? Why use the term at all? If Romanticism reveals "the structure of Goethe's aesthetics and his science, their relationship, and the sources of their construction" (Richards 2002, p. 5) then just use these terms. In any case, I disagree with Richards (2002) regardless of how well-meaning his intentions.[7] Mislabelling something just adds another layer of obfuscation to an area that is already poorly understood.

READING GOETHE'S SCIENTIFIC WORKS

Reading Goethe's scientific works may also confuse the reader in terms of the development of his scientific ideas. Goethe's scientific writings were not published in order of their composition. *An Intermaxillary Bone is Present in the Upper Jaw of Man as Well as in Animals*, for example, was written between 1784 and 1786, well before it was published in 1820. If we are to understand the development of Goethe's method, then we need to read Goethe in order of the year of composition, rather than the year of publication.

In order to measure the development of Goethe's Morphology, namely, Wholeness, Anschauung and the Urphenomenon, we can do this by year of composition and year of publication, using the categories above:

1. Wholeness: a hypothesis (e.g., Urpflanze)
2. Anschauung: observation and measurement (e.g., learning-by-doing)
3. Urphenomenon: recalling the whole

For each part of Morphology we would class these categories as such:

Colour: 1. Polarity; 2. Colour/Turbidity; 3. Urphenomenon
Botany: 1. Urpflanze; 2. Metamorphosis; 3. Urphenomenon
Zoology: 1. Urtier; 2. Comparative Osteology/Metamorphosis; 3. Urphenomenon

Several of the selected works in Table 6.2 are used to show the development of Goethe method in his colour, botany and zoological studies (Figure 6.1).

Goethe conceived the hypothesis of a whole (1 = Wholeness), be it the Urpflanze or Urtier, then observed the phenomena via Anschauung (2), at times using tools (e.g., glass prism) and finally developed the idea, namely the Urphenomenon (3). The development of Goethe's Morphology only makes sense if his works are read in the order of the year of their composition rather than publication.

TABLE 6.2
The Development of Goethe's Method: Morphology

Year of Composition	Year of Publication	Title of Work	Colour	Botany	Zoology
1786–1788	1816	Italian Journey		1	
1790	1790	*The Metamorphosis of Plants*		2	
1791	1791	*Contributions to Optics I*	1		
1792	1792	*Contributions to Optics II*	2		
1795	1891	*Observation on Morphology in General*			1
	1820	*Outline for a general Introduction to Comparative Anatomy, Commencing with Osteology*			2
	1891	*Studies for a Physiology of Plants*		3	
1796–1798	1906?	Metamorphosis of Animals			2
1798	1893	*Empirical Observation and Science*			3
1807	1817	The Purpose Set Forth			3
1808	1810	*Farbenlehre*: Didactic Part	3		

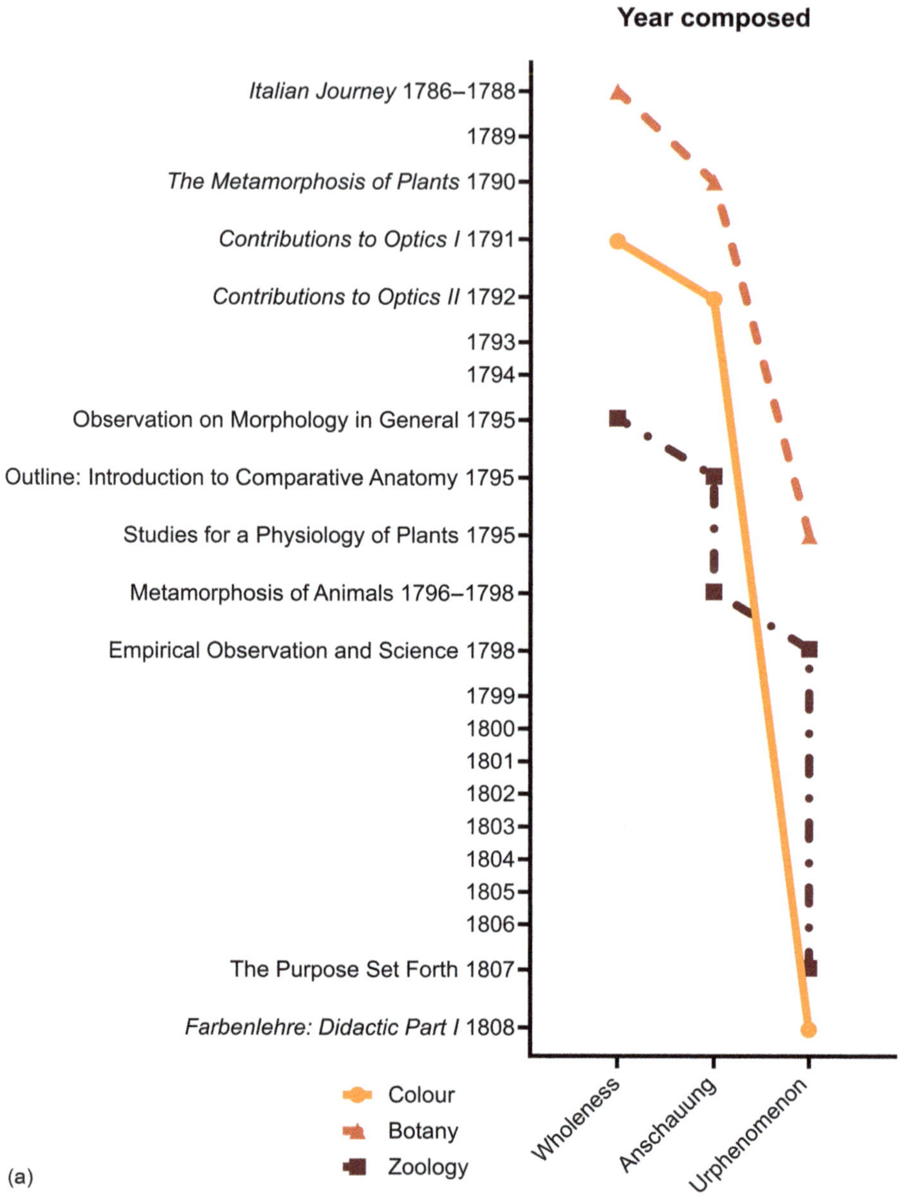

FIGURE 6.1 The development of Goethe's Morphology: Botany (Red), Zoology (Crimson) and Colour (Orange) (1 = Wholeness, 2 = Anschauung and 3 = The Urphenomenon) based on (a) the year composed and (b) on the year published.

(Image by Melinda Tursky, 2026.)

(Continued)

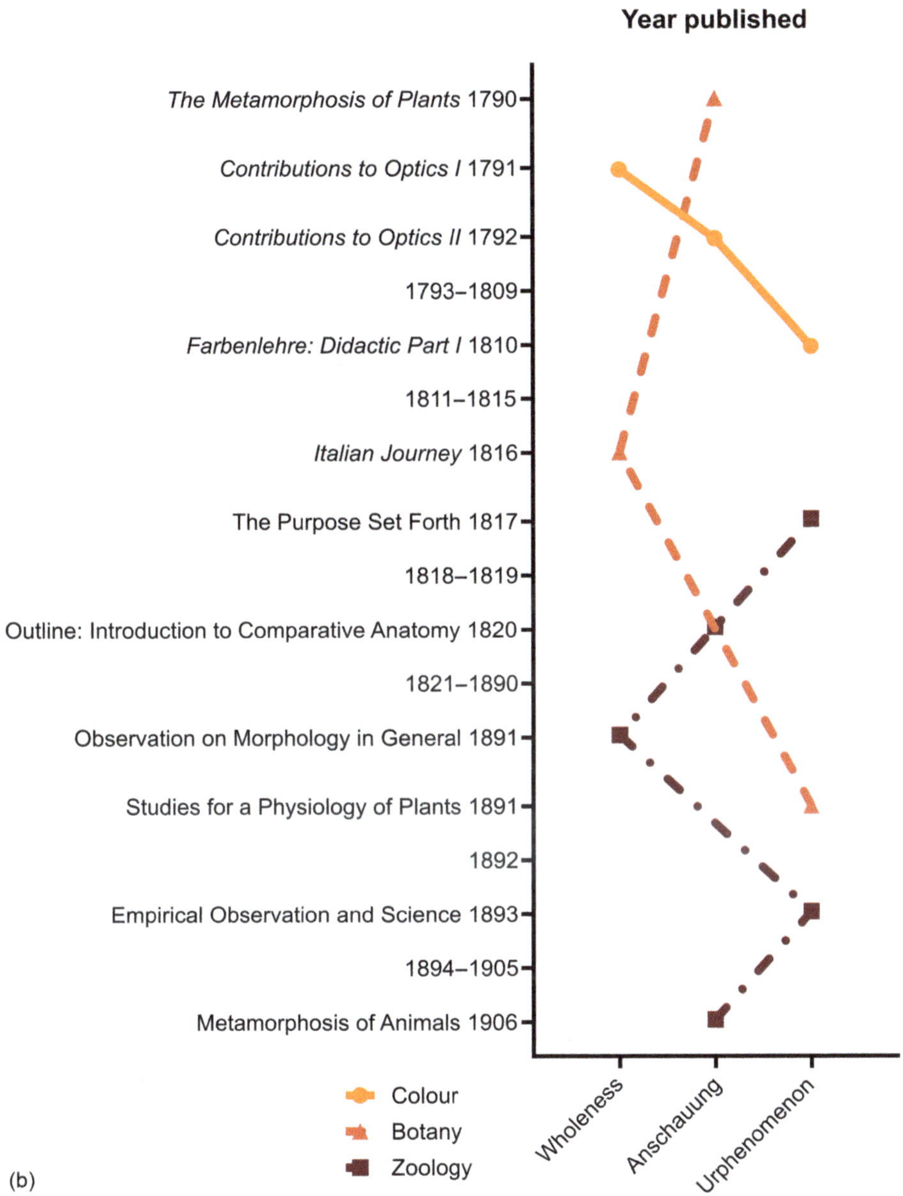

FIGURE 6.1 (Continued)

The development of Goethe's method is most obvious in his works on colour and botany. In his study of colour, Goethe starts off with outlining his hypothesis in 1791 (the first part of *Beiträge*) that prismatic colours form between light and darkness. He uses Anschauung and the glass prism to observe colours on different backgrounds through different turbid mediums, which he achieves by the end of the second part of

Beiträge in 1792. In *Farbenlehre* (1810) he proposes and coins the Urphenomenon. Goethe's method is not obscured in his work on colour theory, if it is read in order by year of composition or by year of publication, namely *Beiträge* followed by *Farbenlehre*. Goethe's method, however, starts to become problematic when reading his botanical work.

Goethe's hypothesis for a whole plant, the Urpflanze, was proposed in his *Italian Journey*, an amalgamation of his diary entries and letters written between 1786 and 1788 and first published in 1816. Goethe's metamorphosis, a learning-by-doing approach, similar to that in his work on colour, was composed and published as the *Metamorphosis of Plants* in 1790. The formation of the Urphenomenon appeared in *Studies for a Physiology of Plants*, published posthumously in 1891. Read by year of publication, Goethe's method would seem to have been created without any hypothesis or aim, as the Urpflanze or any reference to it is missing from the 1790 text.

Goethe's zoological works fair no better when read by year of publication. The Urtier was proposed in 1817 in an introduction to his Morphology. The introduction was written in 1817, but the following work was composed in 1795. The hypothesis of wholeness in animals was proposed in *Observation on Morphology in General*, also composed in 1795. The practise of Anschauung occurred in two works, *Outline for a General Introduction to Comparative Anatomy, Commencing with Osteology* (1795) and a metamorphosis of insects which was written between 1797 and 1798. Goethe's discussion of the Urphenomenon in animals is vague. Goethe mentions the "pure phenomenon" (*sensu* Urphenomenon) in *Empirical Observation and Science*, composed in 1798 and published almost a century later in 1893. There Goethe refers to phenomena he has observed in his "present field of study", without specifying any. Miller suggests that he was working on his *Beiträge* (Miller in Goethe 1995, p. 322), leaving "The Purpose Set Forth", composed in 1807 and published in 1817 as the only other place. Reading Goethe's method in zoology by date of publication would make no sense at all.

Goethe's method is by no means structured as a modern scientific article with an Introduction, Methods, Discussion and Conclusion. We wouldn't expect it to be, given that Goethe was developing his method as he was writing it up. The lack of any formal structure by no means invalidates Goethe's method. Rather it shows us how comparative biology came into being through the application of a method that lies within us all. Remember, Goethe only discovered a method that we as humans have been using for millennia. Goethe started to formalise a method that comparative biologists and taxonomists use every day. The problem lies in working out the development of Goethe's method, because if we read his work by the date of publication, with the exception of his colour theory, his method seemingly disappears, particularly so in his botanical and zoological studies.

GOETHE GETTING IT WRONG

Scientists are not always in possession of all the facts, and much of the time we infer events to make sense of the phenomena around us. There is, however, a rule to inferences: we stick to all possibilities within reason, namely those based on rational arguments regardless of whether they agree or disagree with our worldview. The idea is that we learn from any mistakes or misconceptions and start again. Historical science, into which comparative biology and geology belong, has a lot of data missing. What we try not to do is to favour our reasoning above the absence of data, or worse, use the absence of data as a form of data. If we do this, we place the proverbial horse in front of the cart; something that still plagues modern comparative science.

Geology seems to be where Goethe favoured untested ideas over actual data, namely in his insistence on supporting the Neptunist cause regardless of the evidence. We may say that as an untrained scientist, Goethe simply didn't have the discipline needed to reject Neptunian theory. I do not think that this is the case, as his work in Morphology and *Farbenlehre* demonstrates. Goethe was highly disciplined, but not when it came to geology. In fact, Goethe oscillated between an abstract idea, namely Neptunism, and Anschauung, never discovering the Urphenomenon. His method here seems incongruous:

$$\text{Neptunism} -> \text{Geologising}\left[+\text{Hand lens, Furnace}\right] -> \text{Unobservable Neptunistic events}$$

Goethe noted that without geology his Morphology would never be complete. Unfortunately, Neptunism as an unobservable idea cannot translate into an Urphenomenon. Goethe is just left with an idea (i.e., hypothesis of a whole) but no way of observing it via rocks or minerals. Instead, rocks and minerals are viewed through the distorting lens of a bankrupt hypothesis, leading to speculations and stories rather than anything concrete. Perhaps this is why Goethe wrote,

> What is the inference? Only this, that geology partakes of the uncertainty which pervades every other department of science.
>
> *(LA I 2 p. 151)*

If we were to pick this apart, we may see why Goethe has treated geology as an exception to the rule. "What is inference?" refers to the conclusions we make based on evidence and argumentation. Goethe uses his argumentation to make speculative claims about geological processes *sans* evidence. In the next sentence, "Only this, that geology partakes of the uncertainty which pervades every other department of science", means that geology is an outlier, that it is a historical science like comparative biology, in which we cannot observe past processes and events (*sensu* Wilkins and Ebach 2014). While Goethe restricted his botany and zoology to observable processes in living organisms, he lacked that ability to directly observe the processes that have created the fossil, rock or mineral material. If Goethe had just stuck with inferences and evidence, he would have come close to the origins of granite and basalt. As a rusted-on Neptunist he couldn't drop the ideology and cherry-picked the evidence. The melting and recrystallisation of basalt by Watt and Sosman (see Chapter 2) that validated Hutton's theory should have sent Goethe on the right path, but the millstone of Neptunism simply drowned him.

RECOGNITION

In 1770–1771 Goethe wrote his doctoral thesis, now lost, in jurisprudence, titled "The legislature, on the power of the magistrate to determine religion and culture" [*De legislatoribus, über die Macht der Gesetzgeber, über Religion und Kultus zu bestimmen*], at the University of Strasbourg, which was rejected and suppressed because Goethe outrageously challenged the orthodoxy of the church.

> The dean, a bright, lively man, began praising my work, then proceeded to its questionable aspects, which he by degrees characterised as dangerous, and concluded by saying it might not be advisable publish this work as an academic dissertation.
>
> *(Goethe 1994a, p. 352)*

Goethe instead was awarded a licentiate to practise law on 6 August 1771 (also see Fink 1991, p. 9; Boyle 1991, pp. 103–104).

Goethe's wish for his ideas to be taken seriously fell on deaf ears in Strasbourg, and this was a recurring reaction throughout much of his life. When he discovered the intermaxillary bone, the anatomists Blumenbach and Camper gave Goethe short shrift,

> [w]hen I sent my first discovery of intermediate bones to Peter Camper, I was, to my infinite mortification, utterly ignored. With Blumenbach I fared no better.
>
> *(Goethe to Soret, 2 August 1830)*

In contrast, Goethe's ideas were taken seriously at the esteemed French Academy of Science in the debate between Georges Cuvier (1769–1832) and Étienne Geoffroy Saint-Hilaire (1772–1844) in 1830,

> … you can form no conception of what I felt at the intelligence of the sitting [of the French Academy of Science] on the 19th of July [1830].[8]
>
> *(Goethe to Soret, 2 August 1830)*

Excited by the debate, Goethe reviewed two articles transcribed in *Principes De Philosophie Zoologique* (Saint-Hilaire 1830; see also Appel 1987).

> In earlier times, the meetings of the academy were closed; only the members attended and discussed experiences and opinions. Gradually, friends of science were allowed in as listeners in a friendly manner, and other visitors could not easily be kept out, so that eventually one saw the presence of a significant audience.[9]

In reporting the debate at the academy, Goethe had given the German-speaking public a chance to see what happened in these elite academic circles, thereby opening the debate up to the informed public, something more in line with German natural history societies.

The *Principes De Philosophie Zoologique* (Appel 1987, p. 160) had given Goethe hope that the academy had "changed course", more aligned with the *Meeting of the Natural Research Society, Jena [Tagung der Naturforschenden Gesellschaft, Jena]* that Goethe and Schiller attended almost exactly 36 years earlier on 20 July 1794. The clash between Schiller's synthetic "idea" (i.e., subjective abstraction) and Goethe's analytical idea (i.e., wholeness) must have rekindled a memory. Appel considers that Goethe's interpretation of the debate as one of idea versus facts, or analysis versus synthesis (Appel 1987, p. 161). Goethe saw Geoffroy,

> decidedly on our side, and with him all important scholars and adherents in France. This occurrence is of incredible value to me; I justly rejoice that I have at last witnessed the universal victory of a subject to which I have devoted my life, and which, moreover, is my own *par excellence*.[10]
>
> *(Goethe to Soret, 2 August 1830)*

The Geoffroy–Cuvier Debate gave Goethe much vindication for his science writings. Earlier in November 1792, during his Campaign in France, we see Goethe despair of his scientific legacy, particularly among some of his friends, namely, the family of German philosopher Friedrich "Fritz" Jacobi (1743–1819), during his visit in Pempelfort in Düsseldorf,

> I had little better luck with my nature studies; no one seemed able to understand the passionate earnestness with which I pursued these inquires [...] My friends considered this praiseworthy endeavour a whimsy and a mistake [...] They had taken little notice of my *Metamorphosis of plants*, which had been published a year ago, and when I delivered my thoughts on morphology, no matter how well I knew the subject and no matter how careful I was in its presentation [...] it was all too clear that the unyielding proposition, 'nothing can come into being which does not already exist' had taken firm hold in their minds.

Goethe refers here to the theory of preformation, namely, that the egg and sperm just contain miniature versions of the adult (i.e., homunculi). How could these people, who represented the intellectual class, even begin to understand Goethe's Morphology?

> Something had also been said about my Contributions to Optics [...] for everyone there, no matter how educated, had learned only about the refraction of light, and unfortunately they wanted to reduce everything they saw, even those living things that gave them pleasure, to that of a dead hypothesis.[11]
>
> *(Goethe 1994b, pp. 711–712)*

Goethe would meet Schiller two years later, after which his ideas about Anschauung and the Urphenomenon flourished. Regardless, it is not surprising that Goethe ponders the legacy of his scientific work, given that it has little relevance among his own friends. What of his enduring legacy during his own lifetime? Much of that could be encapsulated in a *Goethean Encyclopedia*.

NOTES

1 The tools in this case were most likely microscopes and other instruments that measured the phenomena. The descriptions in each of these cases are found in their accompanying works.

2 I have been accused of being "mystical" when I presented these ideas to some of my taxonomic colleagues.

3 https://bhma.org/what-is-holism-in-healthcare/

4 … so dürft' es wohl in Intellektuellen derselbe Fall sein, daß wir uns, durch das Anschauen einer immer schaffenden Natur, zu geistigen Teilnahme an ihren Produktionen würdig machte […] Ein lebendig existierendes Ding kann durch nichts gemessen werden was außer ihm ist sondern wenn es ja geschehen sollte müßte es den Maßstab selbst dazu hergeben, dieser aber ist höchst geistig und kann durch die Sinne nicht gefunden werden … (LA I 9, pp. 95–96; LA I II, p. 6)

Original translation by Douglas Miller in Goethe (1995, pp. 31, 8). The original translation uses the terms "spiritually" and "spiritual" when it is clearly "mentally" and "intellectual". Miller also translates "Seele" [soul] into "mind".

5 "Das Klassische nenne ich das Gesunden, und das Romantische das Kranke" (*Gespräche* II, p. 63).

6 "Er bewies mir, daß ich selber, wider Willen, romantisch seiend meine *Iphigenie*, durch das Vorwalten der Empfindung, keineswegs so klassisch und im antiken Sinne sei als man vielleicht glauben möchte" (Gespräche II, p. 137).

7 Fink (1991) also categorises Goethe's scientific writings into "Romantic science" and "Classical science" without clearly defining what these terms mean.

8 … und Sie können sich keinen Begriff machen, was ich bei der Nachricht von der Sitzung des 19. Juli empfinde […] Als ich mein erstes Aperçu vom Zwischenknochen an Peter Camper schickte, ward ich zu meiner innigsten Betrübniß völlig ignoriert. Mit Blumenbach ging es mir nicht besser … (*Gespräche* III, pp. 340–341)

9 In den früheren Zeiten waren die Sitzungen der Akademie geschlossen, nur die Mitglieder fanden sich ein und diskutieren über Erfahrungen und Meinungen. Nach und nach ließ man Freunde der Wissenschaft als Zuhörer freundlich herein, andere Zubringende konnten in der Folge nicht wohl abgehalten werden, und so sah man endlich in Gegenwart eines bedeutenden Publikums. (LA I 10, pp. 400–401)

10 … entscheiden auf unserer Seite und mit ihm alle seine bedeutenden Schüler und Anhänger Frankreichs. Dieses Ereigniß ist für mich von ganz unglaublichem Wert, und ich jubele mit Recht über den endlich erlebten allgemeinen Sieg einer Sache, der ich mein Leben gewidmet habe und die ganz vorzüglich auch die meinige ist. (*Gespräche III*, p. 342)

11 Mit meinem Naturbetrachtung wollte es mir kaum besser glücken; die ernstliche Leidenschaft womit ich diese Geschäft nachhing konnte niemand begreifen […] sie hielten dieses löblichen Bestreben für einen grillenhaften Irrtum […] Von der schon ein Jahr gedruckten Metamorphose der Pflanzen hatten sie wenig Kenntniß genommen, und wenn ich meine morphologischen Gedanken, so geläufig sie mir auch waren, in bester Ordnung und wie es mir schien bis zur kräftigsten Überzeugung vortrug, so mußte ich doch leider bemerken, daß die starre Vorstellungsart: nichts könne werden als was schon sei, sich aller Geister bemächtigt habe. (WA I 33, pp. 196–197)

REFERENCES

Appel TA (1987) *The Cuvier-Geoffroy Debate*. Oxford University Press, Oxford.

Boyle N (1991) *Goethe: The poet and the age (Part 1)*. Oxford University Press, Oxford.

Cantor PA (2004) The scientist and the poet. *The New Atlantis*, 4: 75–85.

Clements T (1949) Leonardo da Vinci as a geologist. *Geological Society of America Bulletin* 60: 1937.

Ebach MC (2015) *Origins of biogeography*. Springer, Dordrecht, Netherlands.

Fink KJ (1991) *Goethe's history of science*. Cambridge University Press, Cambridge.

Goethe JW (1971) *Goethe's colour theory* (Ed. R Matthaei) Studio Vista Ltd, London.

Goethe JW (1994a) *From my life (Parts 1–3)*. The Collected Works 4, Princeton University Press, Princeton.

Goethe JW (1994b) *From my life (Part 4) and Campaign in France/Siege of Mainz*. The Collected Works 4, Princeton University Press, Princeton.

Goethe JW (1995) *Scientific studies* (Vol. 12) (Ed. D Miller). Goethe: The Collected Works, Princeton, New Jersey.

Haeckel E (1876) *The History of Creation* (2nd edition). Henry S. King and Co., London.

Holdrege C (2005) *The Giraffe's long neck*. Nature Institute, Ghent, New York.

Magnus R. (1949) *Goethe as a scientist*. Henry Schuman, New York.

McNamara KJ, Feist R, Ebach MC (2009) Patterns of evolution and extinction in the last Harpetid trilobites during the Late Devonian (Frisian). *Palaeontology*, 52: 11–33.

Medawar P (1968). *The art of the soluble*. Metheun & Co. Ltd.

Merriam-Webster.com Dictionary (2025). Merriam-Webster, "Holism". https://www.merriam-webster.com/dictionary/holism. Accessed 6 October 2025.

Richards R (2002). *The Romantic conception of life*. Chicago University Press, Chicago.

Ritter WE (1932) Why Aristotle invented the word Entelecha. *The Quarterly Review of Biology*, 7: 377–404.

Saint-Hilaire G (1830) *Principes de philosophie zoologique, discutés en mars 1830 au sein de l'Académie royale des sciences*. Pichon et Didier, Paris.

Sherrington C (1949) *Goethe on Nature and on science*. Cambridge University Press, Cambridge.

Smuts JC (1927) *Holism and Evolution*. Macmillan and Company Limited, London.

Steigerwald J (2002) Goethe's morphology: Urphänomene and aesthetic appraisal. *Journal of the History of Biology*, 35: 291–328. https://doi.org/10.1023/A:1016028812658

Wilkins JS, Ebach MC (2014) *The Nature of Classification*. Palgrave Macmillan, New York.

7 Goethean Encyclopaedia

In *Other kindnesses* [*Andere freundlichkeiten*] published in *Zur Morphologie*, Goethe lists a series of grievances in which his work, or natural objects that were named after him, were not recognised by the naturalists of his day. Goethe noted various acknowledgements to his work on botany, such as Jussieu alluding to his *Metamorphosis of Plants*,[1] or the additions to the 1791 edition of Jussieu's *Genera plantarum* added by Paul Usteri (1768–1831),

> On the Metamorphosis of Plants, Goethe has recently written an excellent [egregie] account ...[2]

Goethe lamented that Usteri's additions did not make it into later editions of the *Genera plantarum*. Goethe's work did however receive some praise. The bible of German botany, *Foundations of Botany* [*Grunriß der Kräuterkunde*] (Willdenow 1792), did mention Goethe.

> The life of plants is, as Mr. Goethe very nicely puts it, an expansion and contraction, and these alternations make up the different periods of life.[3]

Willdenow may as well have cited *Faust* for all that it was worth. Goethe wanted to be taken seriously as a botanist, and while Willdenow's words were kind, they were not the nice and authoritative "egregie" of Mr Usteri.

Another loss was the genus *Goethia*, proposed by Batsch (1794, p. 60), that was later not accepted as a valid name.[4] Goethe lamented "I wouldn't know what it might be called now".[5]

"A third attempt to get my name into the annals of science",[6] Goethe notes, was the bone *Goethiano palmigradorum* by Professor Fischer von Waldheim in 1911. Goethe seemed pessimistic, having seen all previous names synonymised or ignored.

> Even this good intention will hardly achieve its purpose, and I will continue to accept seeing such a friendly monument disappear from scientific designations.[7]

Goethe's Bone, as it is known today, is still in use although *Goethiano palmigradorum* has fallen from use.

As with most scientists, Goethe was wondering about his legacy and whether it went beyond plants, minerals and bones dedicated in his name. He acknowledged the various societies that offered him membership (see below), but Goethe still resented his work and ideas being left out of the main scientific literature.

> And should anyone take offence at me for saying all this so openly about myself, should anyone regard such things as inappropriate self-praise, I will soon seize the opportunity to tell just as openly and without reservation how unkindly and repulsively my scientific efforts in a related field have been treated for the past twenty-six years.[8]

DOI: 10.1201/9781003275343-7

Yet there was some acknowledgement,

> For I find that in the most deserving Curt Sprengel's *History of Botany*, just as I am looking through it to gain an overview of the development of such a highly valued science, my work is also remembered with honour. And what greater reward can one imagine than to be approved by such men who one has always had in mind as protagonists in one's undertaking.[9]

The lines in Sprengel that rekindled Goethe's spirit was,

> Goethe presented the comparative development of the plant parts in an extremely clear and engaging way [...] By condensing these forms, the development is prepared: Goethe explains this basic law of vegetation in a convincing and instructive way.[10]

Goethe's remarkable lament, for want of a better word, included his methodology, namely Morphology. He never mentions it by name but refers to it as his "idea" (i.e., practise), one that existed in a world in which there was no interaction between "poetry and science", which "appeared to be the greatest of adversaries".[11]

> As each individual sphere of activity became isolated, the treatment within each circle became more isolated and fragmented. Even a hint of theory [i.e., practise] aroused fear; for over a century it had been avoided like a ghost and, with fragmentary experience, had finally thrown itself into the arms of the most vulgar notions [Vorstellung]. No one wanted to admit that an idea [*Morphology*], a concept that could underlie observation, promote experience, and even encourage discovery and invention.[12]

The disassociation of art and science, famously heralded in 1833 by William Whewell (1797–1866), occurred a year after Goethe's death (see Ebach 2022). Yet Goethe was well aware that it had already happened a century before, around the time of Newton's *Opticks*. Practise, as in Anschauung, and the Urphenomenon were considered to be intuitive and therefore subjective, not something welcome in the increasingly mechanistic science. The "vulgar notions" that Goethe refers to are the idealistic notions of an "objective" science, one that excludes human consciousness and treats observations as subjective and flawed. These notions are enshrined in Newton's *Opticks* as mathematics reveals truths that can only be discovered through logic. Again, we are dragged back into Plato's cave to observe the shadows of natural phenomena that we can never discover through experience alone.

The *Goethean Encyclopaedia* is a self-critique of Goethe's legacy and has been extended herein to include natural objects that were named in his honour (including those after his death in 1832), as well as synonymies of the original names, misnomers (named after people with the same or similar names) and a list of scientific societies that accepted Goethe as a member. A glossary of Goethe's unique scientific terms and a timeline of his publications will also be included.

LEGACY (1829)

The poem *Legacy* [*Vermächtnis*] is found at the end of the second book of *Wilhelm Meister's Journeyman Years*, after a series of 177 Maxims concerning science [Naturwissenschaft]. The last two Maxims are,

> Experience first benefits science, then harms it, because experience makes us aware of law and exception. The average of the two does not give us the truth [176].
>
> One hesitates: between two opposing opinions lies the truth. Not at all! The problem lies in between, the unseeable, the eternally active being thought about peacefully [177].[13]

These are two experiences that are rarely discussed within the scientific method, namely, the problem of two equally valid observations that conflict. Clearly something is going on, an unseen process, which continues to evade our observation and detection. Goethe's geological studies are particularly prone to this phenomenon, particularly in the formation of granites, jointing and basalts (see Chapter 2). Not understanding what these natural processes are must have irritated Goethe, whose own method, Morphology, was mostly descriptive and therefore bombproof (i.e., Goethe's colour exercises still work today, and the intermaxillary bone still features in human anatomy studies). Goethe's poem *Legacy* touches upon several of these points.

Legacy

No being can crumble into nothing,
The eternal moves on in all,
Consider yourself happy!
Being is eternal, for laws
Preserve the living treasures
With which the all is adorned.

The truth found long ago,
Has united noble minds [Geisterschaft[14]],
The old truth, touch it.
Thank it, Earthling [Erdensohn[15]], the wise
Who told it to circle the sun
And showed its siblings the path.

Now turn inwards immediately,
There you will find the centre
Which no nobility can doubt.
You will not miss any rule there,
For independent conscience
Is the sun of your daily customs [Sittentag[16]].

You must then trust your senses,
They will not let you see [schauen] anything false
If your mind[17] keeps you awake.
With fresh eyes joyfully observe,

And mosey [wandle], sure and supple,
Through richly endowed [reichbegabter[18]] meadowy [Auen[19]] world.

Enjoy abundance and blessings in moderation,
Reason is present everywhere
Where life rejoices the living.
Then the past is constant,
The future ahead is alive,
The moment is eternity.

And if you have finally succeeded,
And you are steeped with feeling:
What is fruitful, alone is true;
You scrutinise the general rule,
It will do its own thing,
Join the smallest crowd.

And as from ancient times, in silence,
A work of love, according to their own will,
The philosopher, the poet, create;
So you will gain the most wonderful favour;
To sound out noble souls.
Is the most desirable profession.[20]

CELESTIAL OBJECTS AND CRATERS

- Asteroid 3047 Goethe (6091 P-L) was discovered on 24 September 1960 by Ingrid and Cornelis van Houten-Groeneveld and T. Gehrels. The size of the asteroid (5.846 kilometres in diameter) means it has been reclassified as a minor planet.
- Goethe Basin/Crater, Borealis quadrangle (H-1), Planet Mercury (317 kilometres in diameter).

MOUNTAINS, GLACIERS AND LAKES

- Mount Goethe, Goethe Glacier, Goethe Lake, Central Sierra Nevada Range, John Muir Wilderness, Sierra National Forest, Fresno Co. CA, USA. (Elevation 4045 metres).

  The highest peak of Glacier Divide was named by the U.S. Forest Service and the Sierra Club in 1949, the bicentennial of the birth of Johann Wolfgang von Goethe, poet, philosopher, scientist. The name was suggested by David R. Brower and Erwin G. Gudde and was confirmed by the Geographic Board in August 1949. (Gudde 1960, p. 122)

  Named in 1949 in honor of the German writer and philosopher Johann Wolfgang von Goethe (1749–1832), the author of *Faust*. (Bright 1998, p. 61)
- Mount Goethe, Westland District, West Coast, New Zealand. (Elevation 2350 metres).

To the great German, monuments have been erected here, *aere perennius*: to the Olympian of Weimar, Mount Goethe … (Herz 1912, p. 318–319)

- Goethefelsen [Goethe Rock], Ochsenkopf, Fichtelgebirge, Bavaria, Germany. (Elevation 1020 metres).
- Goethefelsen [Goethe Rock], Bodetal, Harz, Saxony-Anhalt, Germany. (Elevation 205 metres).

  Formerly known as the Siebenbrüderfelsen (Seven Brothers Rock), it was renamed to mark Goethe's 200th birthday on 28 August 1949.
- Goethestein [Goethe Stone], Bodetal, Harz, Sachsen-Anhalt, Germany.

  The Goethe Stone is a granite plate found on the shore of the river Bode, next to the Wanderweg.
- Goethův vrch [Goethe's Rock], Aš ridge, Karlovy Vary Region, Western Bohemia, Czechia. (Elevation 670 m).
- Goethův kámen [Goethe Stone], Aš ridge, Karlovy Vary Region, Western Bohemia, Czechia.
- Goethova skalka [Goethe's Rock Garden], Hazlov, Cheb District, Karlovy Vary Region, Western Bohemia, Czechia.

MINERALS

Goethite [Goethit] is an iron oxide hydroxide coined by Johann Georg Lenz in 1806, from samples found in an iron mine called *Eisenzecher Zug* in Eiserfeld, near Siegen Germany (see Chapter 2).

> … hat man daselbst zu Ehren unseres verhrungswürdigsten Präsidenten, des Herrn Geheimenrath von Göthe obigen Namen beigelegt [… the name above was added there in honor of our most esteemed president, the secret secretary of Göthe].
>
> *(Lenz 1806, p. 46)*

There are two forms of iron oxide hydroxide, the alpha form [α-FeO(OH)] commonly known as Goethite, and the gamma form [γ-FeO(OH)] known as Lepidocrocite (commonly found growing in modern rusty drainage pipes). The mineral proposed by Lenz in consultation with Ludwig Wilhelm Cramer (1755–1832), Reverend Heinrich Adolf Achenbach (1776–1819), Burgermeister [Mayor] of Siegen Johann Daniel Engels (1761–1828) and Friedrich Wilhelm Riemer (1774–1845) refers to the gamma form, which, during Goethe's lifetime, was called Pyrrosiderite (Rubinglimmer), a mineral described in 1814 by Johann Christoph Ullmann (1771–1821).

Unfortunately, Goethite never managed to establish itself as the official term for iron oxide hydroxide because Pyrrosiderite, or rather Lepidocrocite, was used instead. Goethite was reintroduced in 1864 by Franz von Kobell (1803–1882) in his *Geschichte der Mineralogie von 1650–1860* (Kobell 1864).

In 1901 François Alfred Lacroix (1863–1948) concluded that Lepidocrocite had priority as the name for the gamma form but instead assigned the name Goethite to the alpha form [α-Fe^{3+}O(OH)], for which it is known today.

The fossil earwig *Kotejalabis goethitica* Engel & Chatzimanolis 2005 (see below) is named after the unique Goethite preservation.

ROCKS

In 1784 Goethe coined the term granulite to refer to a

> [s]trange rock [he collected] from the Bude Valley, diagonally opposite the Susenburg [Suseburg], which I dare not classify as either granite or porphyry, and for which I suggest the name granulite, because of the quartz grains found in them. It is extremely solid (see above).

Unfortunately, Goethe described what is possibly a kyanite schist, a medium-grade metamorphic rock within the greenschist facies. Weiß (1803) pipped Goethe to the post with his description, which corresponds to granulite found in the Saxonian Granulite complex in the Granulite mountains [Granulitgebirge], near Waldheim, Germany, a place Goethe passed through in August 1797. Goethe's collection also contains another type of metamorphic rock called serpentinite,[21] which was collected by Johann Friedrich Wilhelm von Charpentier (1738–1805) near Waldheim (see Charpentier 1778, p. 68).

ANATOMY

- The Goethe Bone or ossicle, also known as the Inca bone or the preinterparietal bone, was described by Gotthelf Fischer von Waldheim in 1811 as *Goethiano palmigradorum* (Fischer von Waldheim 1811). The bone is located on top of the skull, in the area previously occupied by the fontanelle (Joseph et al. 2022).

ORGANISMS (COMMON NAME OR INDIVIDUALS)

- Goethe's Elephant was an Indian elephant (*Elephas maximus*) that lived between 1771 and 1780 and died in Kassel. Goethe used the specimen to study the intermaxillary bone (see Matuschek 2020).
- Goethe's Amber Ant, a fossil encased in amber from the Eocene (~47–34 Ma), was found in the Goethe's collections. The ant, *Ctenobethylus goepperti*, was described in the late 19th century (see Boudinot et al. 2026).
- Goethe Plant (*Kalanchoe pinnata*[22]), [Brutpflanze, Brutblatt], a succulent plant native to Madagascar, is also known as the Air plant, Floppers, Leaf of Life and Life plant, about which Goethe wrote a poem "With a leaf of *Bryophyllum calycinum*" as well as directions for propagating and growing the plant.[23]
- Goethe's Palm (*Chamaerops humilis*) is commonly known as Mediterranean palm, Dwarf fan palm or European fan palm. The individual located in the Pauda Botanic Garden [*Orto Botanico Università di Padova*] in Italy was planted in 1585 and is also known as Saint Peter's Palm. Goethe saw the plant during his Italian journey on 27 September 1786 but never refers to it. Instead this account appears 45 years later in 1831: "A palm frond sought my full attention …"[24]

- Goethe's Tree (*Cupressus sempervirens*), in the Giardino Giusti, Verona, Italy. "A tree whose branches, from the bottom to the top, from the oldest to the youngest, has taken three hundred years to grow, is certainly worthy of veneration".[25] The tree was almost completely destroyed on 23 August 2020.[26]
- There are several trees named in honour of Goethe.
    - Goethe-Eiche [Goethe Oak, *Quercus robur*] in Silkerode in Thuringia, Germany (felled in 2002[27]); in Marienhöhe, Darmstadt-Eberstadt, Germany; in Arnsberg, Hochsauerlandkreis, North Rhine-Westphalia, Germany; in the Buchenwald Concentration Camp Memorial on Ettersberg hill near Weimar, Saxony, Germany, which was completely destroyed by allied bombing on 24 August 1944.[28]
    - Chêne de Goethe [Goethe Oak] in Sessenheim, Bas-Rhin, France.
    - Goethův dub [Goethe Oak] in Krásný Dvůr, Louny District, Ústí nad Labem Region, Czechia.
    - Goethe-Linde [Goethe Linden Tree, *Tilia cordata*] in Hirschgraben, Frankfurt am Main, Hessen, Germany (felled in 1982[29]); in Wetzlar-Garbenheim, Germany; at Plesse Castle, near Göttingen, Lower Saxony, Germany.
    - Le tilleul de Goethe [Goethe's Linden Tree] between Bastberg and Bouxwiller, Bas-Rhin, France.
    - Goethe-Buche [Goethe Beech, *Fagus sylvatica*] in the Schlosspark Harbke, Harbke, Saxony, Germany[30]; near Frankfurt Airport, Frankfurt am Main, Hesse, Germany.[31]

TAXONOMY

Plants

Goetheaceae[32] Reichb.
Goethia Batsch *(Nomen nudum)*[33]
Goethea Nees synonymised under *Pavonia* Cav.
"… so named after the great German poet, Goethe" (Hooker 1852, tab. 4677).

Goethea alnifolia (A.St.-Hil.) Garcke
Goethea cauliflora (Nees) Fryxell ex G.L.Esteves
Goethea garckeana (Gürke) Garcke
Goethea intermedia (A.St.-Hil.) Heydt
Goethea makoyana (É.Morren) Hook. fil. synonym of *P. makoyana* É.Morren
Goethea multiflora (Juss.) N.E.Br
Goethea semperflorens Nes & Mart synonym of *P. semiserrata* (Schrad.) Steud.
Goethea strictiflora (Hook.) G.L.Esteves

Goethoides Gürke a subgenus of *Pavonia.*
Calycolpus goetheanus (Mart. ex DC) synonym of *Myrtus goetheanus* Mart. ex DC, *Myrtus goetheana* Mart. ex DC and *Campomanesia goetheanus* Hemsl.

Radiolarians

Coccodiscus goethei Haeckel, 1887

Podocyrtis (Lampterium) goetheana (Haeckel, 1887) synonym of *Cycladophora goetheana* Haeckel, 1887.

Haeckeliana goetheana Haeckel, 1887

Porifera (Sponges)

Leucosolenia goethei (Haeckel, 1870) synonym of *Ascaltis goethei* (Haeckel 1870).

Ctenophora (Comb Jellies)

Stenoptycha goetheana (Haeckel 1880)

Dermaptera (Earwigs - Fossil)

†*Kotejalabis goethitica* (Engel and Chatzimanolis 2005)

> The specific epithet is based on goethite [see above], a form of rust, and the mineral by which arthropods of the Crato Member (Santana Formation) are preserved.
>
> *(Engel and Chatzimanolis 2005, p. 224)*

Lepidoptera (Butterflies and Moths)

Papilio forbesi goetheanus Fruhstorfer, 1902 synonymised in to *Papilio forbesi* Grose-Smith, 1883.

Coleoptera (Beetles)

Trichopsida goethei Larochelle and Larivière, 2013

> This species is named after Johann W. von Goethe (1749–1842 [sic]), the German poet, novelist, dramatist, and scientist, who commented on the importance of expertise by writing 'Self-limitation is where you demonstrate you are a master' (Nature and Art, 1800).
>
> *(Larochelle and Larivière 2013, p. 76)*

Hymenoptera (Wasps)

Alexandre Arsène Girault (1884–1941) was a prolific American entomologist who had named over 3000 species. It is herein assumed that *Goetheana* and *Goethella* were named after Goethe even though no etymologies were written for either genus.

Goetheana Girault, 1920

Goetheana shakespearei Girault, 1920 [Type species]

Goetheana incerta Annecke, 1962

Goetheana kobzari Gumovsky, 2016

Goetheana parvipennis (Gahan, 1927)
Goetheana pushkin Triapitsyn, 2005
Goetheana rabelaisi Triapitsyn, 2005
Goetheana thripsivora Narayanan, Subba Rao and Ramachandra Rao, 1960 synonymised under *G. shakespearei* by Bouček, 1988.

Goethella asulcata Girault, 1928
Gonatocerus goethei Girault, 1912

Dedicated to Johann Wolfgang Goethe, poet, naturalist and monistic philosopher.

(Girualt 1912, p. 140)

Eucharomorpha goethei Girault, 1934
Orasemorpha goethei (Girault, 1934)
Mymaromma goethei Girault 1920
Epistenia goethei Girault, 1913
Metapelma goethei Girault, 1928
Ceratoneura goethei (Girault, 1915)
Paraceratoneura goethei Girault, 1915
Parooderella goethei Girault, 1929
Thaumasura goethei Girault, 1937
Anastatus goethei (Girault, 1929)
Gryon goethei (Girault, 1932)

Metatheria (Marsupial-like Fossil)

- †*Amphiperatherium goethei* (Crochet 1979)

Species dedicated to the German writer Goethe who, according to tradition, meditated not far from the deposit of the type species. (Crochet, 1979, p. 371, my translation)

ATTRIBUTIONS TO OTHERS

- Goethe State Forest, Florida, USA, is named after James Tillinghast Goethe (1897–1993).
- Goethe Giant, a cypress tree in the Goethe State Forest, possibly named after J.T. Goethe.
- Species named after George Washington Goethals (1858–1928), US Army General and engineer who oversaw the construction of the Panama Canal: *Halictus goethalsi*, *Goethalsiella*, *Opshomala goethalsi*, *Goethalsia*, *Goethalsia bella* (Goethals' Hummingbird), *Hoplomys goethalsi*, *Teinostoma (Pseudorotella) goethalsi.*
- Species named after Dr Charles Matthias Goethe of Sacramento CA (1875–1966): *Hoplocharax goethei*
- Species possibly named after Belgian chemist Charles van Goethem: *Bathygadus goethemi*, synonymised under *Bathygadus macrops*; *Goethemia pinnulata* synonymised under *Parvicardium pinnulatum*, *Goethemia elegantula*; and the subfamily Goethemiinae.

GOETHE'S MEMBERSHIP OF SCIENTIFIC SOCIETIES AND PRIZES

Goethe was a member and president of various German and international scientific societies, possibly through providing and exchanging mineral and petrological specimens, as well as donating books, manuscripts and other documents on natural history. The societies Goethe participated in included:

- Natural History Society, Jena [*Naturforschende Gesellschaft, Jena*], where he was admitted as an Honorary Member [Ehrenmitglieder] in 1793 (Anonymous 1793, p. 55)
- Society of Mineralogy, Jena [*Sozietät für die gesamte Mineralogie zu Jena*], where he served as President in 1796 (Salomon 1990, p. 73)
- Friday Club [*Freitagsgesellschaft*] in Weimar on 9 September 1791, where "he opened the proceedings with a talk on colour theory" (Boyle 2000, pp. 105–106)
- Moscow Society of Naturalists upon its foundation in 1805 (Mirzoyan 2006, p. 87)
- Prussian Academy of Sciences [*Königlich Preußische Sozietät der Wissenschaften*] on 15 July 1806 as a Foreign member ["auswärtiges Mitglied"] (Harnack 1901 p. 407)
- Foreign member [auswärtiges Mitglied] of the Bavarian Academy of Sciences and Humanities [Bayerische Akademie der Wissenschaften] in 1808 (Willoweit 2009, pp. 23–24)
- The Wetterau Society of Natural History [Wetterauische Gesellschaft für die gesamte Naturkunde zu Hanau] in 1809 (Zingel 1908, pp. 44–46)
- "Goethe was also given the honour of being admitted to the ranks of the Academy [Academia of Science, Erfurt]"[34] on Napoleon's Day, 15 August 1811, (Oergel 1904, p. 221)
- Wernerian Natural History Society [Edinburgh] (Foreign Member in 1811; Anonymous 1811, p. xx)
- Leopoldina National Academy of Science [*Leopoldina Nationale Akademie der Wissenschaften*] on 26 August 1818, as Cognomen Arion IV Diplom (Berg 1991)
- Russian Mineralogical Society (formerly the St. Petersburg Mineralogical Society) in 1818 (Vernadsky 2024, p. 12–13)
- Honorary Member [Ehrenmitglied] of the Silesian Society for German Culture, Breslau [Schlesische Gesellschaft für vaterländische Kultur in Breslau] on 16 July 1822 (Schiff 1928, p. 468)
- In 1823 German botanist Christian Friedrich Schwägrichen (1775–1853) "procured for Goethe the membership of the Royal Society of Edinburgh (then under the presidency of Sir Walter Scott)" (Hennig 1949, p. 542)
- Carl Ludwig Giesecke (1761–1833), "professor of mineralogy at the Royal Dublin Society, was instrumental in having Goethe elected honorary member of the Royal Irish Academy" in 1825 (Hennig 1949, p. 542; also see Waterhouse 1933, p. 214)
- "On the proposal of Count Sergey Semionovich Uvarov (1786–1855) Goethe was elected on 29 December 1826, together with other German, French, English and Italian scientists, as honorary members of the Academy

of Sciences [Russian Academy of Sciences, then St. Petersburg Imperial Academy of Sciences]" (Durylin 1932, p. 211)

- Member of the Naturwissenschaften section of the Göttingen Academy of Sciences and Humanities in Lower Saxony [Niedersächsische Akademie der Wissenschaften zu Göttingen] in 1828 (Arnim 1928, p. 220)
- Goethe is also listed among the participants of the Institute of Archeological Correspondence at the end of the year 1832 as an Honorary member (deceased) [Membri Onorarj] (Anonymous 1832, p. iii), and as a Corresponding member [Membri Corrispondetni] in 1860 of the German Archaeological Institute [originally *Instituto di corrispondenza archeologica*] (Anonymous 1860, p. 9).

Goethe won:

- Presentation award for physics, St Petersburg Academy of Sciences on 28 January 1827 [Physikalische Preis-Aufgabe der Petersburger Akademie der Wissenschaften] (Matthaei 1971, p. 206).

NOTES

1 Goethe must be referring to Antoine Laurent de Jussieu (1748–1836), as his son Adrien Laurent Henri de Jussieu (1797–1853) published his thesis in 1824.

2 "De Metamorphosi Plantarum, egregie nuper Goethe egit, ejus libri analysin uberiorem dabo" (Usteri in Jussieu 1791, p. 487).

3 "Da Leben der Pflanzen ist also, wie Herr Goethe ganz artig sagt, ein Ausdehnen und Zusammenziehen, und jene Abwechslungen machen die verschiedenen Perioden des Lebens aus" (Willdenow 1792, p. 343).

4 Goethe notes that it was not accepted: "… it did not survive the [Linnaean] System" ["Batsch zum Beweise seiner Neigung und Dankbarkeit bildet eine *Goethia* […] sie erhielt sich aber nicht in System"] (LA I 9, p. 103). In 1899 Möbius notes "Thus, Goethe's name has also been preserved externally in botany in the genus *Goethia*, established by Batsch in 1794 among the Scrophulariaceae [i.e., figworts], failed to gain entry into science because the species belonging to it were not listed" ["In dieser Gattung hat sich also Goethes Name auch äußerlich in der Botanik erhalten, nachdem die von Batsch 1794 unter den Scrophulariaceen aufgestellte Gattung *Goethia* keinen Eingang in die Wissenschaft gefunden hatte, weil die hierhergehörigen Arten nicht angegeben waren"] (Möbius 1899, p. 4).

5 "Wie sie jetz heißen mag wüßt' ich nicht anzugeben" (LA I 9, p. 103). I was unable to find any information about *Goethia* after extensive searches within various taxonomic databases.

6 "Einen dritten Versuch meinen Namen in der Wissenschaft ein Denkmal zu setzen …" (LA I 9, p. 104).

7 "Schwerlich wird auch dieser gute Wille seinen Zweck erreichen, und ich werde mir nach wie vor gefallen lassen auch ein so freundliches Denkmal aus den wissenschftlichen Bezeichnungen verschwinden zu sehen" (LA I 9, 104).

8 Und sollte man mir verargen dieses alles ganz unbewunden von mir selbst zu sagen, sollte man dergleichen als ein unziemliches Eigenlob ansehen, so werde nächstens Gelegenheit ergreifen eben so frei und ohne Hinterhalt zu erzählen, wie unfreundlich und widerwärtig man seit sechs und zwanzig Jahren meine wissenschaftlichen Bemühungen in einem verwandten Felde behandelt hat. (LA I 9, p. 104)

9 Denn ich finde, in des verdientesten Curt Sprengels Geschichte der Botanik, eben als ich sie zur Übersicht des Werdens einer so hoch geschätzten Wissenschaft durchschaue, auch meiner Arbeit in Ehren gedacht. und wo kann man sich eine größere Belohnung denken als von solchen Männern gebilligt zu werden, die man bei seinem Unternehmen immer als Protagonisten vor Augen gehabt. (LA I 9, p. 105)

10 "... Goethe die Entwicklung der Pflanzenteile aus einander ungemein klar und einnehmend vor [...] Durch zusammendrägen der Formen wird die Entfaltung vorbereite: diese Grundgesetz der Vegetation führt Göthe auf überzeugende und lehrreiche Art aus" (Sprengel 1818, pp. 302–303).

11 "Kunst und Poesie berührten einander kaum, an lebendige Wechselwirkung war gar nicht zu denken, Poesie und Wissenschaft erschienen als die größten Widersacher" (LA I 9, p. 106).

12 Indem sich nun jeder einzelne Wirkungskreis absonderte, so vereinzelte, zersplitterte sich auch in jedem Kreise die Behandlung. Nur ein Hauch von Theorie erregte schon Furcht; denn seit mehr als einem Jahrhundert hatte man sie wie ein Gespenst geflohen und, bei einer fragmentarischen Erfahrung, sich doch zuletzt den gemeinsten Vorstellung in die Arme geworfen. Niemand wollte gestehen daß eine Idee, ein Begriff der Beobachtung zum Grunde liegen, die Erfahrung befördern, ja das Finden und Erfinden begünstigen könne. (LA I 9, p. 106)

13 Die Erfahrung nutzt erst der Wissenschaft, sodann schadet sie, weil die Erfahrung Gesetz und Ausnahme gewahr werden läßt. Der Durchschnitt von beiden gibt keineswegs das Wahre [176].

Man zagt: Zwischen zwei entgegengesetzten Meinung liege die Wahrheit mitgewinne. Keineswegs! Das Problem liegt dazwischen, das Unschaubare, das ewig tätige Leben in Ruhe gedacht [177]. (LA I 11, 365)

14 "Geisterschaft", as with several other terms in this poem (see "Sittentag" and "reichbegabter"), was possibly invented by Goethe.

15 Matthaei (1971, p. 204) uses the term "Earthling". Literally translated it means "Son of the Earth".

16 Grimm and Grimm define it as "the customary life of man under the image of a day" ["das sittliche leben des menschen unter dem bild eines tages"] https://www.dwds.de/wb/dwb/sittentag accessed: 13 February 2025. In other words, it is what an individual does throughout the day (e.g., walk in the morning, lunch at 1pm and wine with dinner).

17 Interestingly Matthaei translates "Sinnen" as "mind" and "Verstand" as "sense". My translation makes more sense (no pun intended) to Goethe's thinking (the mind cannot "see" as such).

18 The term "reichbegabter" here refers to the meadows.

19 Matthaei (1971, p. 204) has translated this line as "Through eyes on rich-endowed world". The line in WA I 3, p. 82 reads "Durch Auen reichbegabter Welt", not "Durch Augen reichbegabter Welt". This means that there is either the same typographical error in various copies of the poem or Matthaei had read "Augen" [eyes] instead of "Auen" [meadow]. Given that I have not seen any other copy of the poem use the term "Augen" (see Goethe 1981, pp. 195–197, 1982, p. 337, 1994, pp. 266–269), I will use the term "meadow" instead.

20 Vermächtniß.

Kein Wesen kann zu nichts zerfallen,
Das Ew'ge regt sich fort in allen,
Am Sein erhalte dich beglückt!
Das Sein ist ewig, denn Gesetze
Bewahren die lebend'gen Schätze
Aus welchen sich das All geschmückt.

Das Wahre war schon längst gefunden,
Hat edle Geisterschaft verbunden,
Das alte Wahre fass' es an.
Verdank' es, Erdensohn, dem Weisen
Der ihr die Sonne zu umkreisen
Und dem Geschwister wies die Bahn.

Sofort nun wende dich nach innen,
Das Centrum findest du da drinnen
Woran kein Edler zweifeln mag.
Wirst keine Regel da vermissen,
Denn das selbstständige Gewissen
Ist Sonne deinem Sittentag.

Den Sinnen hast du dann zu trauen,
Kein Falsches lassen sie dich schauen
Wenn dein Verstand dich wach erhält.
Mit frischem Blick bemerke freudig,
Und wandle, sicher wie geschmeidig,
Durch Auen reichbegabter Welt.

Genieße mäßig Füll' und Segen,
Vernunft sei überall zugegen
Wo Leben sich des Lebens freut.
Dann ist Vergangenheit beständig,
Das Künftige voraus lebendig,
Der Augenblick ist Ewigkeit.

Und war es endlich dir gelungen,
Und bist du vom Gefühl durchdrungen:
Was fruchtbar ist allein ist wahr;
Du prüfst das allgemeine Walten,
Es wird nach seiner Weise schalten,
Geselle dich zur kleinsten Schaar.

Und wie von Alters her, im stillen,
Ein Liebewerk, nach eignem Willen,
Der Philosoph, der Dichter schuf;
So wirst du schönste Gunst erzielen;
Denn edlen Seelen vorzufühlen
Ist wünschenswertester Beruf

(WA I 3, pp. 82–83)

21 Rock specimen No. 2848 in Catalogue A (A3 Suite von Kursachsen) "Serpentinite. From Waldheim [Serpentin. Von Waldhiem]" (Prescher 1978, p. 161).

22 Synonymies include *Bryophyllum pinnatum*.

23 "Mit einem Blatt *Bryophyllum calycinum*" (WA I 4, p. 276); for propagating and growing *Bryophyllum calycinum*, see LA I 10, pp. 211–213, 228.

24 "Eine Fächerpalme zog meine ganze Aufmerksamkeit auf sich …" (LA I 10, p. 333).

25 "Ein Baum, dessen Zweige von unten bis oben, die ältesten wie die jüngsten, gen Himmel streben, der seine dreihundert Jahre dauert, ist wohl der Verehrung Werth" (WA I 30, p. 75).

26 https://www.visitverona.it/en/poi/goethe-s-tree-575
27 https://www.monumentale-eichen.de/abgestorbene-eichen/silkerode/
28 Previously known as the "Fat Oak" ["Dicke Eiche"] https://www.buchenwald.de/en/geschichte/historischer-ort/konzentrationslager/goethe-eiche
29 Known as the "Old Lime Tree" ["Die alte Linde"] (Maisak and Pravida 2017, p. 265, f. 71).
30 "A present from Johann Wolfgang Goethe 1805" ["Ein Geschenk des Johann Wolfgang von Goethe 1805"], https://www.monumentaltrees.com/en/deu/saxonyanhalt/borde/10696_schlossparkharbke/20907/
31 "Goethebuche" https://www.monumentaltrees.com/en/deu/hesse/frankfurt/21448_flughafens/
32 "Barnhart (1895) and Moldenke (1944) name is cited in synonym. – *Goethea* Nees, 1821. – Barnhart (1895, p. 7) incorrectly attributed family to rank to not a validly published [...] name" (Hoogland and Reveal 2005, p. 98). In other words, Goetheaceae is not recognised as valid name and is synonymised into *Goethea.*
33 The following description accompanied the name "h) calyx 7-partite, 4 opposite subulate lobes, 3 alternate ovate, larger; corolla infundibuliform, 5-partite; stamens didyma; stigma obtuse, emarginate; capsule 2-locular, polyspermous 511. GOETHIA" ["h) calyce 7- partito, laciniis 4 oppositis subulatis, 3. alternis ovatis, majoribus; corolla infundibuliformi, 5- partita; staminibus didynamis; stigmate obtuso, emarginato; capsula 2- loculari, polysperma 511. GOETHIA"] (Batsch 1794, p. 60). A modern interpretation of the description would be: "Sepals in 7 parts; 4 of these long elongated and tapering to a point, 3 are alternate to the others, oval shaped and larger; Petals in 5 parts, funnel shaped; stamens 4 of equal length; stigma rounded and notched; fruit a capsule with two spaces and many seeds" (Murphy pers. comm. 2025). In the 2018 *International Code of Nomenclature for algae, fungi, and plants*, Nomen nudum (or "naked name") is "[a] designation of a new taxon published without a description or diagnosis or reference to a description or diagnosis (Art. 38 Ex. 1, Rec. 50B)" (https://www.iapt-taxon.org/nomen/pages/main/glossary.html?zoom_highlight=nomen+nudum).
34 "... dass auch Goethe die Ehre zuteil wurde, in die Zahl der Akademie [Akademie gemeinnütziger Wissenschaften zu Erfurt] aufgenommen zu werden" (Oergel 1904, p. 221).

REFERENCES

Annecke DP (1962) The genus Goetheana Girault in South Africa (Eulophidae-Hymenoptera). *South African Journal ofAgricultural Science*, 5(2): 273–279.

Anonymous (1793) *Nachricht von der Gründung einer Naturforschenden Gesellschaft zu Jena.* Cunoischen Buchhandlung, Jena.

Anonymous (1811) *Memoirs of the Wernerian Natural History Society Vol. 1 for the years 1808, 1809, 1810*, White & Cochran, and J. Murray, London.

Anonymous (1832) Elenco de' Partecipanti. *Bullettino dell'Instituto di corrispondenza archeologica*, 1: i–vi.

Anonymous (1860) Elenco de' Partecipanti. *Bullettino dell'Instituto di corrispondenza archeologica*, 32: 1–11.

Arnim M (1928) *Mitglieder-Verzeichnisse der Gesellschaft der Wissenschaften zu Göttingen (1751–1927).* W. Fr. Kaestner, Göttingen.

Barnhart JH (1895) Family nomenclature. *Bulletin of the Torrey Botanical Club*, 22: 1–24.

Batsch AJGK (1794) *Synopsis universalis analytica generum plantarum* (vol. 2). Croeker, Jena.

Berg W (1991) Arion IV. Goethe als Mitglied der Leopoldina. In: Salve Academicum II. Beiträge zur Geschichte der Deutschen Akademie der Naturforscher Leopoldina. Schweinfurt. *Veröffentlichungen des Stadtarchivs Schweinfurt*, 5: 109–126.

Bouček Z (1988) *Australasian Chalcidoidea (Hymenoptera): A biosystematic revision of fourteen families, with areclassification of species.* CAB International, Wallingford, UK

Boudinot BE, Bock BL, Tröger D, Weingardt M, Hammel JU, Grabe V, Solórzano-Kraemer MM, Jandausch K, Oberski JT, Schmuck T (2026). Discovery of Goethe's amber ant: Its phylogenetic and evolutionary implications. *Scientific Reports* 16: 2880.

Boyle N (2000) *Goethe: The poet and the age (Vol. 2)*. Oxford University Press, Oxford.

Bright W (1998) *1500 California place names: Their origin and meaning, a revised version of 1000 California Place Names by Erwin G. Gudde*, 3rd Edition. University of California Press, Berkeley.

Charpentier JFW von (1778) *Mineralogische Geographie der chursächsischen Lande*. Siegfried Lebrecht Crusius, Leipzig.

Crochet J-Y (1979) Diversité systématique des Didelphidae (Marsupialia) européens tertiaires. *Géobios* 12:365–378

Durylin C (1932) Дурылин С. Русские писатели у Гете в Веймаре/Предисловие редколлегии – С.81-504. Литературное наследство [И.В. Гёте]/Том 4/6//Редколлегия: Л.Авербах, И.Ипполит (Ситковский), Ф.Рас кольников; Ответственный редактор: И.Ипполит; Заведующий редакцией: И.Зильберштейн/План тома, литературная редакция, подбор иллюстраций, оформление И.Зильберштейна, С.Макашина, И.Сергиевского; Обложка работы И.Ф.Рерберга. – На суперобложке.: Goethe./ Москва: Журнально-газетное объединение ("Жургаз"), 1932. – 1 062 с., ил. – 10 000 экз.

Ebach MC (2022) *Goethe in the Age of Artificial Intelligence*. Palgrave Macmillan, Singapore.

Engel MS, Chatzimanolis S (2005) Early Cretaceous earwigs (Dermaptera) from the Santana Formation, Brazil. *Polskie Pismo Entomologiczne* 74: 219–226.

Fischer von Waldheim GF (1811) Observata queadam de osse epactali sive Goethiano palmigradorum. Prodromo inservientia craniologiae comparatae. Anatomicis celeberrimis Blumenbach, Cuvier, Soemmerring, dicata a Gothelf-Fischer, Philosophiae et Medicinae Doctore, Professore Mosquensi, Ordinis Sti. *Vladimiri quartae classis Equite, etc. Cum tabulis tribus aeri incisis*. Typis Nicolai Sergeidis Vsevolojsky, Moscow.

Fruhstorfer H (1902) Neue Papilioformen und andere Lepidopteren and Ost-Asien und dem malayischen Archipeligo. *Deutsche Entomologische Zeitschrift Iris*, 15: 306–315.

Gahan AB (1927) Miscellaneous descriptions of new parasitic Hymenoptera with some synonymical notes. *Proceedings of the United States National Museum*, 71: 1–39.

Girault AA (1912) Australian Hymenoptera Chalcidoidea. II. The family Mymaridae with descriptions of new species. *Memoirs of the Queensland Museum*, 1: 117–175.

Girault AA (1913) More new genera and species of chalcidoid Hymenoptera. Hymenoptera from Paraguay. *Archiv für Naturgeschichte*, 79: 51–69.

Girault AA (1915) Australian Hymenoptera Chalcidoidea IV. *Supplement. JOURBOOK: Memoirs of the Queensland Museum*, 3: 180–299.

Girault AA (1920) New genera and species of Australian Mymaridae (Hymenoptera). *Insecutor Inscitiæ Menstruus*, 8: 96–100.

Girault AA (1928) *Some Insecta and a new All Highness (Notes compiled in fear and sorrow)*. Published privately, Brisbane.

Girault AA (1929) Notes on, and descriptions of, chalcid wasps in South Australian Museum. Concluding paper. *Transactions of the Royal Society of South Australia*, 53: 309–346.

Girault AA (1932) *New lower Hymenoptera from Australia and India*. Private Publication, Brisbane.

Girault AA (1934) *Eucharitidae, Cynipidae, Proctotrypidaeet Thysanoptera Nova Australiensis*. Privately Published, Brisbane.

Girault AA (1937) *New naturals unorthodoxies and non-pollutions viz new hexapods*. Privately published, Brisbane.

Goethe JW (1981) *Goethe*. Selected Poems. Heinemann Educational, London.

Goethe JW (1982) *Wilhelm Meisters Wanderjahre*. Reclam, Stuttgart.

Goethe JW (1994) *Selected poems*. Princeton University Press, New Jersey.

Grose-Smith H (1883) Descriptions of three new species of *Papilio. The Entomologists monthly Magazine* 18: 234–235.

Gudde EG (1960) *California place names: The origin and etymology of current geographical names.* University of California Press, Berkeley.

Gumovsky A (2016) Review of Afrotropical species of Goetheana Girault (Hymenoptera: Eulophidae), with description of a new species. *Zootaxa* 4147: 551–563. http://doi.org/10.11646/zootaxa.4147.5

Haeckel E (1870) Prodromus of a system of the Calcareous sponges. *Annals and Magazine of Natural History*, 5: 176–191.

Haeckel E (1880) Das System der Acraspeden. 2te Hälfte des Systems der Medusen. Acht Nachträge zur Vervollständigung des Systems. *Denkschriften der Medicinisch-Naturwissenschaftlichen Gesellschaft zu Jena.* 2: 361–672.

Haeckel E (1887) Report on the Radiolaria collected by H.M.S. Challenger during the years 1873–76. Reports of Science Research Voyage of H.M.S. Challenger. *Zoology*, 32, 1–92.

Harnack A (1901) *Geschichte der Königlich Preussischen Akademie der Wissenschaften zu Berlin : Ausgabe in einem Bande / im Auftrage der Akademie bearbeitet von Adolf Harnack.* Verlag von Georg Stilke, Berlin.

Hennig J (1949) Goethe's interest in British mineralogy. *Proceedings Linnean Society London*, 161, 99–207.

Herz M (1912) *New Zealand: The country and the people.* Duffield and Co., New York.

Hoogland RD, Reveal JL (2005) Index Nominum Familiarum Plantarum Vascularium. *Botanical Review*, 71: 1–291.

Hooker WJ (1852) *Curtis's botanical magazine*, Vol. LXXVIII, Reeve and Co., London.

Joseph S, Mathew AJ, Sukumaran TT, Pavikuttan N, Palatty BU, Sudhakaran R (2022) The Inca Bone: A throwback to an ancient civilization and people. *Journal of Morphological Sciences*, 39: 115–118.

Jussieu AL (1791) Genera plantarum Secundum Ordines Naturales Disposita, *Juxta Methodum in Horto Regio Parisiensi Exaratam. Ziegleri et Filigorum*, Zurich.

Kobell F (1864) *Geschichte der Mineralogie von 1650–1860.* Cottaschen Buchhandlung, Munich.

Larochelle A, Larivière M-C (2013) Carabidae (Insecta: Coleoptera): synopsis of species, Cicindelinae to Trechinae (in part). *Fauna of New Zealand* 69: 1–193.

Lenz JG (1806) *Tabellen über das gesammte Mineralreich.* Göpferdts, Jena.

Maisak P, Pravida D (2017) Das Goethe-Haus in Jahr 1900. *Jahrbuch des Freien Deutschen Hochstifts*, 2017: 246–294.

Matthaei R (1971) *Goethe's colour theory.* Studio Vista Ltd, London.

Matuschek O (2020) *Goethes Elefanten. Insel-Bücherei, Nr. 1489.* Insel Verlag, Berlin.

Mirzoyan EN (2006) Two Hundred Years of the Moscow Society of Naturalists. *Herald of the Russian Academy of Sciences*, 76: 85–88.

Möbius M (1899) Goethe als Botaniker. *Gartenwelt*, 35: 1–8.

Moldenke HN (1944) *A preliminary classification of the plant kingdom to families*, pp. 1–37. New York Botanical Garden, New York.

Narayanan ES, Subba Rao BR, Ramachandra Rao M (1960). Some new species of Chalcids from. India. *Proceedings of the National Academy of Sciences, India Section B: Biological Sciences* 26: 168–175.

Oergel D (1904) Die Akudemie nützlicher Wissenschaften zu Erfurt von ihrer Wiederbelebung durch Dalberg bis zu ihrer endgültigen Anerkennung durch die Krone Preulien (1776–1816). *Jahrbücher der Königlichen Akademie gemeinnütziger Wissenschaften zu Erfurt*, 30: 139–224.

Prescher H (1978) *Goethes Sammlungen zur Mineralogie, Geologie und Paläontologie: Katalog.* Akademie-Verlag, Berlin.

Salomon, J (1990) Die Sozietät für die gesamte Mineralogie zu Jena unter Goethe und Johann Georg Lenz. *Mitteldeutsche Forschungen*, 98. Böhlau, Köln.

Schiff J (1928) Der alte Goethe und Schlesien. *Schlesische Monatshefte: Blätter für Kultur und Schrifttum der Heimat*, 11: 466–471.

Sprengel KPJ (1818) *Kurt Sprengels Geschichte der Botanik.* Vol. 2. F.A. Brockhaus, Leipzig.

Triapitsyn SV (2005) Revision of Ceranisus and the related thrips-attacking entedonine genera (Hymenoptera: Eulophidae) ofthe world. *African Invertebrates*, 46: 261–315.

Vernadsky V (2024) Thoughts and observations on Goethe as a Naturalist. *Publications of the English Goethe Society*, 93(2): 148–199. https://doi.org/10.1080/09593683.2024.2348294

Waterhouse G (1933) Goethe, Giesecke, and Dublin. *Proceedings of the Royal Irish Academy: Archaeology, Culture, History, Literature*, 41: 210–218.

Weiß CS (1803) Über die Gebirgsarten des sächsischen Erzgebirges, ein Schreiben an Herrn Geh, *Ober-Bergrath Karsten mit einigen Anmerkungen von dem 'Letztern Der Gesellschaft Naturforschender Freunde zu Berlin Neue Schriften*, 4: 342–366.

Willdenow C L (1792) *Grundriss der Kräuterkunde*. Haude and Spener, Berlin.

Willoweit D (2009) *Denker, Forscher und Entdecker: eine Geschichte der Bayerischen Akademie der Wissenschaften in historischen Portraits*. CH Beck, Munich.

Zingel J (1908) *Geschichte der Wetterauischen Gesellschaft für die gesamte Naturkunde : eine Festgabe zur Feier ihres 100 jährigen Bestehens*. Wetterauischen Gesellschaft, Hanau.

Index

Pages in *italics* refer to figures, pages in **bold** refer to tables, and pages followed by n refer to notes.

M

N

O

P

R

S

T

For Product Safety Concerns and Information please contact our EU representative GPSR@taylorandfrancis.com
Taylor & Francis Verlag GmbH, Kaufingerstraße 24, 80331 München, Germany

www.ingramcontent.com/pod-product-compliance
Ingram Content Group UK Ltd.
Pitfield, Milton Keynes, MK11 3LW, UK
UKHW022211190926
475702UK00003B/302
* 9 7 8 1 0 3 2 2 3 0 3 8 2 *